# SpringerBriefs in Public Health

**Series Editor**
Macey Leigh Henderson
Georgetown University Kennedy Institute of Ethics
Washington, District of Columbia, USA

More information about this series at http://www.springer.com/series/10138

SpringerBriefs in Public Health present concise summaries of cutting-edge research and practical applications from across the entire field of public health, with contributions from medicine, bioethics, health economics, public policy, biostatistics, and sociology.

The focus of the series is to highlight current topics in public health of interest to a global audience, including health care policy; social determinants of health; health issues in developing countries; new research methods; chronic and infectious disease epidemics; and innovative health interventions.

Featuring compact volumes of 50 to 125 pages, the series covers a range of content from professional to academic. Possible volumes in the series may consist of timely reports of state-of-the art analytical techniques, reports from the field, snapshots of hot and/or emerging topics, elaborated theses, literature reviews, and in-depth case studies. Both solicited and unsolicited manuscripts are considered for publication in this series.

Briefs are published as part of Springer's eBook collection, with millions of users worldwide. In addition, Briefs are available for individual print and electronic purchase.

Briefs are characterized by fast, global electronic dissemination, standard publishing contracts, easy-to-use manuscript preparation and formatting guidelines, and expedited production schedules. We aim for publication 8-12 weeks after acceptance.

Heather Mullins-Owens

# Integrative Health Services

Ethics, Law, and Policy for the New Public
Health Workforce

 Springer

Heather Mullins-Owens
Director of Integrative Health
HERO Network LLC
Indianapolis, IN, USA

ISSN 2197-1935                          ISSN 2197-1943    (electronic)
SpringerBriefs in Public Health
ISBN 978-3-319-29855-9          ISBN 978-3-319-29857-3    (eBook)
DOI 10.1007/978-3-319-29857-3

Library of Congress Control Number: 2016932887

Printed on acid-free paper

This Springer imprint is published by Springer Nature
The registered company is Springer International Publishing AG Switzerland

# Contents

# Chapter 1
# Introduction to Concepts of Integrative Health Services

**Abstract** This chapter introduces definitions and concepts related to integrative health services and complementary and alternative medical practices. A brief overview of the increasing popularity of complementary and alternative medical practices is presented and recent trends are reviewed. The allopathic health care model will be discussed and contrasted with holistic health approaches, and the integrative medical model will be presented.

**Keywords** Health services • Integrative health services • Integrative medicine • Complementary and Alternative Medicine (CAM) • Biomedical model • Integrative medical model • Allopathic medicine • Public health

## 1.1 Student Learning Objectives

After reading this chapter, you should be able to:

• Differentiate between complementary, alternative, and integrative medicine
• Identify strengths and benefits of both the allopathic approach and the more holistic approaches
• Understand the prevalence of CAM within the U.S. health system
• Become familiar with the principals of integrative medicine

## 1.2 Definitions and Concepts

Complementary and Alternative Medicine (CAM) refers broadly to practices and providers that originate 'outside of mainstream medicine' (National Center for Complementary and Integrative Health (NCCIH), 2014). Complementary Medicine refers more specifically to therapies utilized in conjunction with allopathic health care (NCCIH, 2014). Alternative Medicine refers to therapies used in place of allopathic health care, and is less common in the United States than complementary therapies (Bauer-Wu, Ruggie, & Russell, 2009; NCCIH, 2014). Thus, the modalities of both of these types of medicine are generally the same; it is primarily the utilization by the patient that differs. However, use of the term "CAM" itself is the source of great disagreement within the profession.

In recent years, the term "integrative health" has gained traction as an alternative to "CAM." Integrative health is distinguished from CAM in that it denotes a coordination of care which typically includes both allopathic and complementary practices, attempting to provide patients with the best of both. Integrative health applies a holistic model of medicine that stresses wellness of the entire person (physical, mental, social, and often spiritual dimensions of wellness are considered), and places an emphasis on providing an effective, collaborative team of health providers who emphasize patient-centered care and shared decision making between providers and patients. Integrative health and medicine "seeks to build a bridge between conventional and alternative medical systems and to find therapeutic and cost effective ways to combine them so as to have the 'best of both worlds' while still maintaining the integrity of each system" (Maizes, Rakel, & Niemiec, 2009, p. 6.). Often, complementary care providers favor the term *integrative health*, and physicians working in an integrative model favor the term *integrative medicine*, this text will use the terms integrative health and integrative medicine interchangeably.

## 1.3  Increased Popularity of CAM and Integrative Health and Medicine

CAM, integrative health and integrative medicine have become increasingly popular in recent decades, and nearly 40 % of adults are now using CAM modalities (Mayo Clinic, 2011). One of the greatest challenges to integrative health has been the way in which medical billing is performed in the United States. Because many complementary therapies do not qualify for reimbursement by insurance companies at the point of service, patients receiving complementary treatments often pay for them out of pocket instead. This keeps integrative health and CAM out of the reach of many patients who may most benefit from the practices.

CAM has become increasingly popular in the United States as patients seek new ways to cope with chronic diseases and symptoms (Barnes, Bloom, & Nahin, 2008a; Barnes, Powell-Griner, McFann, & Nahin, 2004; Frenkel & Arye, 2001). CAM encompasses a great range of interventions and modalities, including Traditional Chinese Medicine (TCM), Ayurveda, homeopathy, acupuncture, reiki, massage and yoga therapy (Maizes et al., 2009). Studies have repeatedly shown the effectiveness of some modalities of CAM such as yoga therapy and acupuncture for various conditions, which has led to some physicians incorporating them into their medical clinics. Often this merging happens under the umbrella term *integrative medicine*.

As early as 1992, estimates showed that the number of visits to CAM providers exceeded the number of visits to all primary care physicians (Frenkel, Ben-Arye, Geva, & Klein, 2007; Wolsko, Eisenberg, Davis, Ettner, & Phillips, 2002). Data from 1997 shows that this increased to a ratio of 1.6, which equaled 629 million visits to CAM providers (Wolsko et al., 2002). As these numbers show, CAM has been popular in addressing problems that traditionally brought people into primary care practices—such as arthritis and joint pain, depression, back and neck pain, headaches, fibromyalgia, stress, and ADD/ADHD (Barnes et al., 2008a). Data from both 2002 and 2007 show that CAM is most popular among adults ages 30–69, and among those with higher levels of education (Barnes et al., 2008b).

One recent study showed that 75 % of CAM visits annually are attributable to 8.9 % of the population (roughly 17,500,000 adults) (Wolsko et al., 2002). However, this type of ratio is similar to that seen in other care providers where a small number of patients account for a majority of provider visits (Wolsko et al., 2002). Insurance coverage for a given CAM modality is strongly correlated with high-frequency use (Wolsko et al., 2002). This correlation is particularly true for manipulative therapies such as chiropractic care and massage therapy (Wolsko et al., 2002). This correlation has not yet been explained. For some it may not be affordable to receive regular CAM treatments without the assistance of insurance coverage. Others may not have considered trying a CAM treatment if it had not been offered in the insurance plan. Selection bias is another [possible explanation for the high numbers of CAM visits among few patients] (people preferred plans offering CAM coverage because they wanted to seek these types of therapy), as is moral hazard increased.

Studies have repeatedly shown the effectiveness of some modalities of CAM such as yoga and acupuncture for various conditions, which has led to some physicians incorporating them into their medical practice. Often this merging happens under the umbrella term *integrative medicine*. Integrative medicine could be defined as a whole-person approach to medical treatment of the mind, body and spirit which supports a more collaborative care approach between patients and providers.

Some innovative integrative medicine clinics are now bringing primary care physicians and CAM practitioners together and merging the practices, although it is unknown how many of these clinics currently exist across the country (Frenkel et al., 2007). Some studies show support among patients of integrating CAM and primary care together so that a physician may oversee the care patients are receiving from alternative practitioners (Frenkel et al., 2007). While some clinics such as this appear very successful (i.e., the Arizona Center for Integrative Medicine), others struggle to find appropriate ways to bill patients for services not covered by most insurance companies, or for visits with physicians that may take much longer than the customary time for appropriate billing codes. After taking a look at the benefits of integrative care for patients (which include higher quality care, greater patient satisfaction, and better symptom management), this paper will discuss how integrative medicine could be more widely incorporated into the current health services delivery system to provide greater efficiency and improved care for patients while keeping costs for chronic conditions sustainable.

## 1.4   Public Health Impacts: Integrative Health and Medicine

Integrative medicine gained popularity as more and more health consumers and providers recognize that high-tech medicine is not adequately addressing chronic disease, health promotion, or disease prevention (Maizes et al., 2009). Most traditional allopathic medical care is heavily focused on acute care which accounts for a much smaller percentage of patients in recent decades (Maizes et al., 2009). Today, more than 133 million Americans live with chronic disease, and 70 % of deaths are caused by chronic disease (Maizes et al., 2009).

Chronic conditions may be defined as a condition that requires ongoing interaction with the health care system. Health interactions for chronic conditions are focused on managing the disease or symptoms rather than curing the condition. If the condition lasts more than 3 months, it meets the U.S. National Center for Health Statistics definition of a chronic disease (National Center for Health Statistics. Health, United States, 2014). Common chronic diseases include arthritis, diabetes, cardiovascular disease, and obesity. The cost of chronic disease care is more than $1.5 trillion annually, and it accounts for 75 % of all medical expenses in the United States (Maizes et al., 2009). Nearly half of all Americans are living with a chronic condition, and 88 % of Americans 65 or older live with a chronic disease (Partnership for Solutions, 2004).

Since many chronic conditions are manageable by medical experts but not curable, the traditional allopathic models of health are not always as effective with these conditions. Greater use of integrative health services may potentially help make managing these conditions more affordable for the already strained U.S. health care system. An integrative approach may also provide other benefits for the health system, such improved preventative health approaches, reduced reliance on expensive medications, even improved provider-patient communication.

## 1.5   The Integrative Medical Model

"Integrative medicine emphasizes the therapeutic relationship between practitioner and patient, is informed by evidence, and makes use of all appropriate therapies" (Arizona Center for Integrative Medicine, 2014). Dr. Andrew Weil, one of the first proponents of integrative medicine in the U.S., founded the Arizona Center for Integrative Medicine (ACIM) in 1994, and serves as its director. ACIM (2014) provides and follows eight principles of integrative medicine:

1. Patient and practitioner are partners in the healing process
2. All factors that influence health, wellness, and disease are taken into consideration, including mind, spirit, and community, as well as the body
3. Appropriate use of both conventional and alternative methods facilitates the body's innate healing response

4. Effective interventions that are natural and less invasive should be used whenever possible
5. Integrative medicine neither rejects conventional medicine nor accepts alternative therapies uncritically
6. Good medicine is based in good science. It is inquiry-driven and open to new paradigms
7. Alongside the concept of treatment, the broader concepts of health promotion and the prevention of illness are paramount
8. Practitioners of integrative medicine should exemplify its principles and commit themselves to self-exploration and self-development (ACIM, 2014).

The integrative medical model is characterized by a collaborative approach between physicians, patients, and various care providers. Traditionally the provider-patient relationship is viewed as a patient seeking the advice of various practitioners with the only commonality being the patient. All communication in this traditional model flowing from one provider to another typically flows through the patient, and as the research suggests very little information may flow at all (Maizes et al., 2009). An integrative medical model is characterized as a dynamic, collaborative model with providers discussing the patient's care with each other and with the patient. This model seems to present each provider with a more complete picture of the patient's conditions, needs, and abilities, as well as the patient's health goals and/or priorities. The Integrative Medical Model also allows the providers to benefit each other with knowledge gained in their specialty.

The importance of both prevention and intervention in determining an appropriate care plan for patients is a key component of the integrative model (Arizona Center for Integrative Medicine, 2014). In the United States, mainstream health care is delivered via the intervention model of care. That is to say, it is delivered upon patient complaint of symptoms, or upon discovering a problem through medical testing and diagnostics (routine or otherwise). There has been a gradual shift towards ensuring that Americans receive some basic preventative care, however, the intervention model is still primary (U.S. Department of Health and Human Services, 2015).

A prevention-based model focuses more attention on ways of ensuring optimal health, rather than simply combating known disease or acute problems. Due to the high cost of intervention for many diseases, more money will likely always spend more on intervention than prevention. However, greater effort can ensure that more medical interactions have a focus on prevention and health optimization. Health practitioners may also find ways to collaborate with public health organizations and nonprofits to develop and implement programs that optimize health in their communities.

By working to promote health rather than wait for disease development, complementary modalities can address early stages of the disease process that may not be identified or addressed with allopathic care (Maizes et al., 2009). Patients are considered well when their body, mind, and spirit are in a relative state of harmony. In the context of integrative health, providers tend to view health as a dynamic condition impacted on a daily basis by a variety of personal, psychological, environmental, cultural, and medical factors. According to the Life Course Health Development Framework:

> Health is a consequence of multiple determinants operating in nested genetic, biological, behavioral, social and economic contexts that change as a person develops; and health development is an adaptive process composed of multiple transactions between these contexts and the biobehavioral regulatory systems that define human functions. (Halfon & Hochstein, 2002, p. 433)

By considering lifespan health development, some proponents of the Life Course Health Development Framework argue that risk factors and early life experiences that impact health later in life might be better anticipated and the negative effects managed or minimized. According to Maizes et al. (2009) the primary barriers to implementing these types of health promoting medical practices are frequently cited as the following:

- Not enough time to spend with patients;
- Little economic incentive for a health focused practice;
- Limited resources to help patients make lifestyle and behavior changes;
- Limited collaboration between clinicians and administration; and
- Need for cultural shift among physicians to a team based knowledge sharing model (p. 282).

## 1.6   Integrative Medicine in Practice

Currently, the most common model in which patients participate in complementary care is in a patient-directed model, where patients take the initiative to supplement their medical care and seek out their complementary care providers (Maizes et al., 2009). Patients maintain all control and must coordinate their own care among all of their allopathic and complementary medical providers (Maizes et al., 2009). There are several concerns with this model. First, patients may not always disclose the complementary therapies they are using with all of their other medical providers. In some instances, particularly with homeopathy, this can create concerns of potential drug interaction. This concern could be addressed either by increased questioning by allopathic providers about complementary therapy usage, or by training pharmacists to inquire about homeopathic use.

Second, this model leaves patients without a primary care physician or health team to make referrals to appropriate and effective practitioners. Just as some allopathic providers are better trained and more familiar with particular patient populations, the same is true for complementary providers. In fact, it may be even more true of complementary providers given that some modalities do not require strict licensing and regulation in the same way that some allopathic fields do.

Pioneering integrative medical centers are becoming more common with the increasing numbers of patients interested in this health approach, and it is especially popular in cancer centers. M.D. Anderson, Sloan-Kettering, and Dana-Farber have been pioneers in this health care niche. It is likely that cancer centers have taken the lead in innovative care due to the larger amounts of financial support they receive from former patients to support such programs. Many other hospitals also offer CAM therapies that fall short of integrative medical care. In a 2010 American Hospital Association (2011) study, 42 % of responding hospitals stated that one or more CAM therapies were being offered (up from 37 % in 2007, and from 26.5 in 2005 American Hospital Association, 2011; Maizes et al., 2009). Eight-five percent of the hospitals stated patient demand was the primary reason for the offering of these services, and 70 % of survey respondents cited clinical effectiveness as their top reason (AHA, 2011). Seventy-five percent of hospitals cited finances as the largest obstacle to implementation of more CAM programs (AHA, 2011).

## 1.7  Questions for Discussion

1. After learning more about the integrative medical model, do you think the health care you typically receive fits within the integrative model? If not, which aspects of the integrative model are missing?
2. What advantages do you see in the integrative medical model?
3. Which patient populations do you think would most benefit from an integrative health services approach?

## 1.8  Definitions

| | |
|---|---|
| Allopathic Medicine | medical system focused on use of remedies such as drugs or surgeries to alleviate or manage symptoms or disease |
| Complementary and Alternative Medicine (CAM) | refers generally to all medical practices deemed to be outside of the mainstream standard of care |
| Complementary Medicine | therapies that are not part of allopathic medicine but are used in conjunction with allopathic treatments |
| Alternative Medicine | therapies not a part of allopathic care and used in place of allopathic interventions |
| Integrative Health | denotes a coordination of care or intervention choices considers use of both allo- |

| | pathic and complementary practices, attempting to provide patients with the best of both therapies |
| Integrative medicine | integrative medicine is a whole-person approach to medical treatment of the mind, body and spirit, which supports a more collaborative care approach between patients and often multiple health providers |
| Chronic conditions | ailments that require ongoing interaction with the health care system. Frequently such ailments are manageable but not curable. |
| Integrative medical model | characterized by a collaborative approach between physicians, patients, and various care providers |
| Preventive Care | health services that focus on avoiding illness, disease, and related problems through services such as health screenings, testing, education, check-ups and patient counseling. |

# References

American Hospital Association. (2011, September 11). More hospitals offering complementary and alternative medicine services. Press Facts regarding the 2010 Complementary and Alternative Medicine Survey. Retrieved from file:///C:/Users/hlmullin/Downloads/110907-pr-camsurvey.pdf

Arizona Center for Integrative Medicine. (2014). *What is integrative medicine?* http://integrative-medicine.arizona.edu/about/definition.html

Barnes, P. M., Powell-Griner, E., McFann, K., & Nahin, R. L. (2004, June). Complementary and alternative medicine use among adults: United States, 2002. In *Seminars in Integrative Medicine* (Vol. 2, No. 2, pp. 54–71). Philadelphia, PA: WB Saunders.

Barnes, P. M., Bloom, B., & Nahin, R. L. (2008a). *National Health Statistics Reports*. Hyattsville, MD: US Department of Health and Human Services.

Barnes, P. M., Bloom, B., & Nahin, R. L. (2008b). *Complementary and alternative medicine use among adults and children: United States, 2007.*

Bauer-Wu, S., Ruggie, M., & Russell, M. (2009). *Communicating with the Public about Integrative Medicine*. Institute of Medicine commissioned paper.

Frenkel, M., & Arye, E. B. (2001). The growing need to teach about complementary and alternative medicine: Questions and challenges. *Academic Medicine, 76*(3), 251–254.

Frenkel, M., Ben-Arye, E., Geva, H., & Klein, A. (2007). Educating CAM practitioners about integrative medicine: An approach to overcoming the communication gap with conventional health care practitioners. *Journal of Alternative and Complementary Medicine, 13*(3), 387–392.

Halfon, N., & Hochstein, M. (2002). Life course health development: An integrated framework for developing health, policy, and research. *Milbank Quarterly, 80*(3), 433–479.

Maizes, V., Rakel, D., & Niemiec, C. (2009). Integrative medicine and patient-centered care. *Explore, 5*(5), 277–289.

Mayo Clinic. (2011). Retrieved from http://www.mayoclinic.org/alternative-medicine/ART-20045267?p=1

National Center for Complementary and Integrative Health (NCCIH). (2014). *Complementary, alternative, or integrative health: What's in a name?* Retrieved from https://nccih.nih.gov/health/integrative-health

National Health Council (November 2, 2015). *About chronic conditions*. Retrieved from: http://www.nationalhealthcouncil.org/newsroom/about-chronic-conditions.

Partnership for Solutions. (2004, September). *Chronic conditions: Making the case for ongoing care*. Retrieved from http://www.partnershipforsolutions.org/DMS/files/chronicbook2004.pdf

U.S. Department of Health and Human Services. (2015, September 22). *Preventive services covered under the Affordable Care Act. Fact Sheets*. Retrieved from http://www.hhs.gov/healthcare/facts/factsheets/2010/07/preventive-services-list.html.

Wolsko, P. M., Eisenberg, D. M., Davis, R. B., Ettner, S. L., & Phillips, R. S. (2002). Insurance coverage, medical conditions, and visits to alternative medicine providers: Results of a national survey. *Archives of Internal Medicine, 162*(3), 281–287.

# Chapter 2
# Application to Public Health

**Abstract** This chapter will discuss how integrative health services and holistic approaches to health can improve overall patient health and health systems outcomes. This will include discussions of improvements in preventative care, chronic disease management, increasing health care costs, and improving system efficiency. The chapter will conclude with a discussion of end of life care and the recent trend of Physician Orders for Life-Sustaining Treatment.

**Keywords** Chronic disease • Integrative medicine • Integrative health • Complementary and Alternative Medicine (CAM) • Chronic care model • Chronic illness • Preventive care • Health spending • End of life • Physician Orders for Life-Sustaining Treatment POLST

## 2.1 Student Learning Objectives

After reading this chapter, you should be able to:

* Summarize the challenges chronic disease present in the current health system
* Identify three areas in which preventive care can save health care dollars.
* Identify the time of life in which most patients incur the most medical costs
* Explain the key components of the Chronic Care Model
* Describe when, how, and for which patients POLST orders may be a valuable tool

This chapter will discuss how integrative health services and holistic approaches to health can improve overall patient health and health systems outcomes. This will include discussions of improvements in preventative care, chronic disease management, increasing health care costs, and improving system efficiency. The chapter will conclude with a discussion of end of life care and the recent trend of Physician Orders for Life-Sustaining Treatment.

© Springer International Publishing Switzerland 2016
H. Mullins-Owens, *Integrative Health Services*,
SpringerBriefs in Public Health, DOI 10.1007/978-3-319-29857-3_2

## 2.2   Chronic Disease Management

Today, nearly half of all Americans are living with a chronic condition (Partnership for Solutions, 2004). Chronic conditions may be defined as a condition that requires ongoing interaction with the health care system. Health interactions for chronic conditions are focused on managing the disease or symptoms rather than curing the condition. If the condition lasts more than 3 months, it meets the U.S. National Center for Health Statistics definition of a chronic disease. Common examples of chronic diseases include arthritis, diabetes, cardiovascular disease, and obesity. The chronic care model promotes interventions that encourage increased self-management of health ailments and symptoms (Wagner et al., 2001). The model indicates productive interactions taking place between informed and engaged patients and prepared and proactive health teams (Wagner et al., 2001). Further, it shows these actions as evolving and flowing from a community which supports self-management of disease symptoms and from a health system that offers decision support and useful clinical information systems.

## 2.3   Preventive Care

The benefits of preventive care are numerous given the prevalence of preventable chronic disease in the health system. It is estimated that 70 % of the health care costs burden is related to preventable illness and disease (Fries et al., 1993). When placing the nine most common causes of death under reclassification according to their underlying actual causes, eight of the nine leading categories were preventable causes (Fries et al., 1993). Furthermore, lifetime medical costs are also clearly linked to higher and lower risk health behaviors (Fries et al., 1993). Evidence also shows that education about health decisions lowers the costs of long term care even in chronic disease patients (Fries et al., 1993). Literature also supports that well designed workplace health promotion programs are able to reduce the illness burden both in terms of direct healthcare costs and in terms of the number of sick days workers require (Fries et al., 1993).

The inability of the U.S. healthcare system to meet demands of chronic conditions is at least in substantial part due to a poorly organized delivery system (Wagner et al., 2001). Improvements in care must come from changing the systems of care, since the system is ill equipped to handle the type of long term health conditions that plague our system (Institute of Medicine, 2001). The Affordable Care Act (ACA) signed into law in 2010 offers some improvements to the medical system by requiring insurance companies to provide several health screenings and preventive services without copays or coinsurance to increase the likelihood consumers seek these services (U.S. Department of Health & Human Services, 2015). The ACA also provides employer incentives to provide more robust workplace wellness programs.

## 2.4   Costs of Care

It is thought by many integrative experts that integrative health care can lower costs by increasing efficiency and effectiveness and minimizing duplication and waste (Kodner & Spreeuwenberg, 2002). This is due to the combining of the inputs, delivery and management of services in the integrative model (Kodner & Spreeuwenberg, 2002). According to Guarneri, Horrigan, and Pechura (2010), cost savings could quickly be established by providing integrative lifestyle change programs for chronic disease patients along with other preventive strategies to promote wellness in other populations (p. 308).

In addition to systemic cost savings, integrative approaches also save money by decreasing utilization of expensive high tech medical interventions and replacing them with lower cost but similarly effective treatments (Guarneri et al., 2010). One example supported by research is the use of acupuncture treatments as an alternative for knee surgery, which has been shown to save $9000 on average (Christensen et al., 1992). Acupuncture has also been shown to produce an average savings of $26,000 by reducing time spent in hospitals or nursing homes after a stroke (Johansson, Lindgren, Widner, Wiklund, & Johansson, 1993).

The complementary and integrative approach to treatment tends to lower costs by providing many mind-body therapeutic options in a small group setting and lowering patient reliance on medications (Guarneri et al., 2010). Group settings not only reduce costs, but have been shown to also improve health outcomes and increase patient satisfaction (Maizes, Rakel, & Niemiec, 2009). These group visits could involve health education, exercise, mindfulness/meditation instruction, or a variety of other options for patients (Maizes et al., 2009). Typically these group classes are arranged by patient population (Maizes et al., 2009).

Finally, integrative medicine can lower health care costs by investing in preventative health. It is less expensive to prevent disease than to have to treat it (Guarneri et al., 2010), particularly considering that most modern diseases are chronic conditions that continue over long periods of time. Serxner, Gold, Meraz, and Gray (2009) reviewed 120 studies of employee wellness programs that resulted in the employers saving 26 % in health care costs. This resulted in an average return on investment of $5.81 for every $1 spent on workplace wellness (Serxner et al., 2009).

## 2.5   End-of-Life Care

End of life care presents an unusual burden on the health care system in terms of economic costs. It is estimated that 18 % of lifetime costs for medical care are incurred during the last year of life (Fries et al., 1993). Even more surprising perhaps is that 29.4 % of Medicare and Medicaid payments for those over 65 years of age are incurred during the last year of life (Fries et al., 1993).

With 70 % of people expressing a preference against life-sustaining treatments at end of life, much of the expense not in the self-expressed interests of many patients (Fries et al., 1993). However, given that only 9 % of the population executes a living will or advance directive, these patient preferences are often not followed by physicians providing end of life care (Fries et al., 1993). Providing the aging population affordable and easy access to executing advance directives may improve patient care, increase patient autonomy, afford humane and dignified care for patients who prefer to avoid particular measures, and reduce health costs. Unfortunately, even when patients make the effort to execute advance directives, they are frequently not available when needed, not transferred with the patient, are not specific enough to be honored, or are otherwise not executed properly (Dunn, Tolle, Moss, & Black, 2007, p. 33).

In order to address these traditional concerns, many states are now providing standardized forms that some state legislatures have named Physician Orders for Life-Sustaining Treatment (POLST). The POLST forms originated in Oregon in 1991, but are now available in some form throughout most of the country (currently it is not available in Alaska, Alabama, Arkansas, Nebraska, or South Dakota) (POLST, 2012–2015, Programs in your state). Physicians complete these forms with seriously ill patients[1] about their end of life care preferences (Dunn et al., 2007). Proponents of these programs argue that they reduce medical errors, better clarify and identify patient wishes, and ensure meaningful discussion and shared decision-making between patients and their physicians. (The National Physician Orders for Life-sustaining Treatment Paradigm, 2015, What is POLST?). The decisions are to be made with a physician examining the patient, and in accordance with the patient's preferences, values, personal beliefs, and taking into account their current state of health. Putting this information at the front of the patient's medical record, and sending a copy home with patients (if appropriate) helps to ensure the information is available and recognizable as a valid physician order to any health care workers treating the patient (Dunn et al., 2007).

## 2.6  Questions for Discussion

1. Do you think POLST will make end of life care decisions easier for patients and their families? Why or why not?
2. What wellness programs or opportunities does your university or workplace offer? Have you participated in any? What was your experience like?

---

[1]As a rule of thumb, it is recommended that providers ask themselves "Would I be surprised if this person died in the next year?" If the answer is No, they would not be surprised, then a discussion of POLST with the patient is recommended (Dunn et al., 2007, p. 37).

## 2.6.1   Case Study #1

Joe is a 66 year old male in generally good condition. He was diagnosed with diabetes 10 years ago and has had high blood pressure for most of his life. Joe is now at the local hospital to receive treatment for a broken arm after falling from a ladder while performing repairs on his home. Is Joe a good candidate for POLST? Why or why not?

## 2.6.2   Case Study #2

Gina is a 65 year old female who was diagnosed 1 year ago with Stage IV ovarian cancer. She was informed her condition is terminal, and is undergoing chemotherapy to try to prolong her life. Her oncologist expects her life expectancy to be in the 6–18 month time frame, although survival time is difficult to predict. She is now in the hospital and it appears she suffered a minor stroke. Is Gina a good candidate for POLST? Why or why not?

## 2.7   Definitions

| | |
|---|---|
| Preventive Care | health services that focus on avoiding illness, disease, and related problems through services such as health screenings, testing, education, check-ups and patient counseling. |
| Chronic Care Model | Framework for improving care for the chronically ill at the individual and population level by identifying six fundamental areas that support the proper management of chronic illnesses. |
| Physician Orders for Life-Sustaining Treatment | advance care planning facilitated between health providers and patients to determine what if any life-sustaining treatments are desired by seriously ill patients in accordance with their preferences, values, personal beliefs, and current state of |

health. Sometimes referred to
by other names depending on
the jurisdiction, such as Medical
Orders for Life-Sustaining
Treatment (MOLST); or
Physician's Orders on Scope of
Treatment (POST).

# References

Christensen, B. V., Iuhl, I. U., Vilbek, H., Bülow, H. H., Dreijer, N. C., & Rasmussen, H. F. (1992). Acupuncture treatment of severe knee osteoarthrosis. A long-term study. *Acta Anaesthesiologica Scandinavica, 36*(6), 519–525.
Dunn, P. M., Tolle, S. W., Moss, A. H., & Black, J. S. (2007). The POLST paradigm: Respecting the wishes of patients and families. *Annals of Long Term Care, 15*(9), 33.
Fries, J. F., Koop, C. E., Beadle, C. E., Cooper, P. P., England, M. J., Greaves, R. F., … Wright, D. (1993). Reducing health care costs by reducing the need and demand for medical services. *New England Journal of Medicine, 329*(5), 321–325.
Guarneri, E. M., Horrigan, B. J., & Pechura, C. M. (2010). The efficacy and cost effectiveness of integrative medicine: A review of the medical and corporate literature. *Explore, 6*(5), 308–312.
Institute of Medicine. (2001). *Crossing the quality chasm: A new health system for the 21st century*. Retrieved from https://iom.nationalacademies.org/~/media/Files/Report%20Files/2001/Crossing-the-Quality-Chasm/Quality%20Chasm%202001%20%20report%20brief.pdf.
Johansson, K., Lindgren, I., Widner, H., Wiklund, I., & Johansson, B. B. (1993). Can sensory stimulation improve the functional outcome in stroke patients? *Neurology, 43*(11), 2189–2192.
Kodner, D. L., & Spreeuwenberg, C. (2002). Integrated care: Meaning, logic, applications, and implications—A discussion paper. *International journal of integrated care, 2*.
Maizes, V., Rakel, D., & Niemiec, C. (2009). Integrative medicine and patient-centered care. *Explore, 5*(5), 277–289.
Partnership for Solutions. (2004, September). *Chronic conditions: Making the case for ongoing care*. Retrieved from http://www.partnershipforsolutions.org/DMS/files/chronicbook2004.pdf
Serxner, S., Gold, D., Meraz, A., & Gray, A. (2009). Do employee health management programs work. *American Journal of Health Promotion, 23*(4), 1–8.
The National Physician Orders for Life Sustaining Treatment Paradigm. (2012–2015). *Programs in your state*. Retrieved from http://www.polst.org/programs-in-your-state/
The National Physician Orders for Life-sustaining Treatment Paradigm. (2015). *What is POLST?* Retrieved from http://www.polst.org/about-the-national-polst-paradigm/what-is-polst/
U.S. Department of Health and Human Services. (2015, September 22). *Preventive services covered under the Affordable Care Act. Fact Sheets*. Retrieved from http://www.hhs.gov/health-care/facts/factsheets/2010/07/preventive-services-list.html.
Wagner, E. H., Austin, B. T., Davis, C., Hindmarsh, M., Schaefer, J., & Bonomi, A. (2001). Improving chronic illness care: Translating evidence into action. *Health Affairs, 20*(6), 64–78.

# Chapter 3
# Models of Health and Health Theory: Focus on Integrative Models

**Abstract** Models of health help us by providing a conceptual framework to the way we interpret and investigate health and illness. Some models, urge us to consider primarily physical and biological aspects of wellness and illness, while others incorporate or even focus on social, spiritual, emotional, and other aspects of wellness and illness. Integrative medical practitioners tend to prefer models that focus on a variety of factors or dimensions of health. The most popular and widely embraced tend to focus, at a minimum, on biological, social and psychological dimensions of health. Models which also incorporate spiritual dimensions of health are also gaining popularity in both clinical practice and in the medical literature. The models in this chapter (other than the biomedical model which they are contrasted with) are all consistent with the principles of integrative health principles and will help us expand our views of health and wellness.

**Keywords** Models of health • Health models • Spiritual dimensions of health • Integrative health • Social determinants of health • Holistic health • Biopsychosocial • Wellness

## 3.1 Student Learning Objectives

After reading this chapter, you should be able to:

- Describe the strengths and weaknesses of the biomedical model
- Understand the most prominent medical models and theories that health professionals use to guide their decision making
- List examples of the social determinants of health

Models of health help us by providing a conceptual framework to the way we interpret and investigate health and illness. Some models, urge us to consider primarily physical and biological aspects of wellness and illness, while others incorporate or even focus on social, spiritual, emotional, and other aspects of wellness and illness. Integrative medical practitioners tend to prefer models that focus on a variety of factors or dimensions of health. The most popular and widely embraced tend

© Springer International Publishing Switzerland 2016
H. Mullins-Owens, *Integrative Health Services*,
SpringerBriefs in Public Health, DOI 10.1007/978-3-319-29857-3_3

to focus, at a minimum, on biological, social and psychological dimensions of health. Models which also incorporate spiritual dimensions of health are also gaining popularity in both clinical practice and in the medical literature. The models in this chapter are all consistent with the principles of integrative health principles and will help us expand our views of health and wellness.

## 3.2  The Biomedical Model

Western medicine was dominated by the biomedical model of health for more than a century. According to this reductionist model, "all disease is a product of a biologic defect often initiated by a biologic pathogen (Johnson, 2012)." This model proved popular and successful when people were dying largely of infectious diseases as they were at the turn of the twentieth century, when the average life expectancy was 47 years (Johnson, 2012). By the end of the twentieth century the leading cause of death was chronic disease, and the average life expectancy was 77 years, leaving many health experts looking for a better model to explain health, and account for the diseases the biomedical model could not easily explain (Johnson, 2012).

The biomedical model focuses heavily on diagnosis and treatment. Diagnosis can be defined as the identification of illness or disease through observation or diagnostics (tests). Treatments may refer to any actions taken or prescribed by a health professional. Examples of the application of the biomedical model include decisions to gather more information through x-rays, blood tests, ultrasounds, and MRIs. Additional examples include prescription medications, and surgeries.

While the biomedical model's effectiveness at diagnosing and treating many conditions should be noted, it also has some shortcomings. First, it relies heavily on professional health providers and high technology equipment and tests, which increase health care costs. Second, not all conditions can be successfully treated. Third, it does not promote health by encouraging healthy habits and lifestyles. Finally, in terms of its application to integrative health, the biomedical model views the body and mind as separate entities and focuses on treating the body without consideration of the role psychology, stress, and emotions may play on health (Borrell-Carrió, Suchman, & Epstein, 2004).

Promotion of healthy habits and lifestyles is important because so many chronic diseases that burden our health care system are substantially attributable to lifestyle choices. The World Health Organization for many years has defined *health* as "a complete state of physical, social and mental wellbeing, and not merely the absence of disease and infirmity (WHO, 1946)." The biomedical model is largely silent on issues of social and mental wellbeing. Furthermore, the biomedical model has been far less effective at diagnosing and treating many modern chronic diseases than it was at diagnosing and treating more acute illnesses. This has led for many in the health community to begin looking to other health models to also apply when diagnosing and treating patients.

## 3.3 The Biopsychosocial Model

The biopsychosocial model was proposed by George Engel in 1977. Although Engel did not disagree with everything the biomedical model offered, he wanted to offer a more humanistic model that empowered patients (Borrell-Carrió et al., 2004). According to the biopsychosocial model biological, psychological and social factors all impact patient health and illness (Engel, 1977). Today, most health experts argue this biopsychosocial approach is better able to explain and potentially the high rates of chronic disease in the U.S. (Borrell-Carrió et al., 2004). In the last two decades, an increased acceptance of this model led to an increased focus on patient-provider communication, disease prevention and management through life-style choices, the impact of culture and beliefs, and increasingly inter-disciplinary health provider teams (Johnson, 2012).

The model puts health and wellness in a more contextual and individualized set-ting (Borrell-Carrió et al., 2004). It also views health as a complex and dynamic state, rather than linear as indicated in the biomedical model (Borrell-Carrió et al., 2004).

## 3.4 The Holistic Model of Wellness

The holistic model of wellness was developed by Bill Hettler, and focuses on six dimensions of wellness (Gieck & Olsen, 2007). These are typically described as: physical wellness, emotional wellness, intellectual wellness, spiritual wellness, occupational wellness, and social wellness (Gieck & Olsen, 2007). This holistic model is often applied to Eastern and complementary medical therapies. By focus-ing on the whole of the individual, this model seeks prevent illness in the body as much as it does to cure disease, and treatments focus on fixing the cause of disease or illness rather than just treating the symptoms (Gieck & Olsen, 2007; Witmer & Sweeney, 1992).

## 3.5 Social Cognitive Theory

According to social cognitive theory, behavior is determined by expectancies and incentives. Individuals make decisions as they weigh the interplay of internal fac-tors (such as knowledge, skills, emotions, and habits) and environmental factors (such as social approval, physical environment, and organizational rules). Three primary factors that impact human behavior are expected to be self-efficacy, goals, and outcome expectancies.

## 3.6    The Health Belief Model

G. M Hochbaum's Health Belief Model (1958) was developed to explain why some people did not participate in public health screening programs in the 1950s (Hochbaum, 1958). According to the health belief model, "health-related action depends upon the simultaneous occurrence of three classes of factors:

1. The existence of sufficient motivation (or health concern) to make health issues salient or relevant
2. The belief that one is susceptible (vulnerable) to a serious health problem...
3. The belief that following a particular health recommendation would be beneficial in reducing the perceived threat, and at a subjectively-acceptable cost. Cost refers to perceived barriers that must be overcome in order to follow the health recommendation; it includes, but is not restricted to, financial outlays. (Rosenstock, Strecher, & Becker, 1988)."

The health belief model is typically used to explain health behaviors and design treatment programs. The six constructs include perceived susceptibility; perceived severity; perceived benefits; perceived barriers; cues to action; and self-efficacy. (Retrieved from http://www.jblearning.com/samples/0763743836/chapter%204. pdf) (Fig. 3.1) Hayden, J. A. (2013).

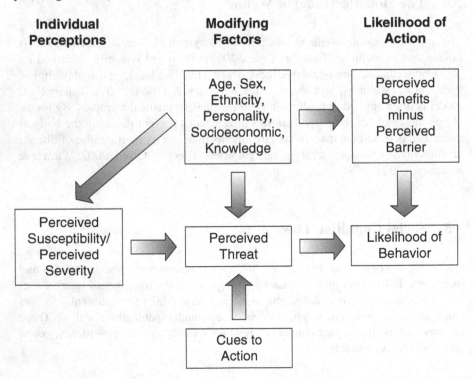

**Fig. 3.1** Health belief model (ch. 4). Hayden, J. A. (2013). Introduction to health behavior theory. Jones & Bartlett Publishers. Permissions granted

## 3.7 The Social Model and Social Determinants of Health

The social model of health considers the social, cultural, economic and environmental factors of health instead of focusing on disease and illness like the biomedical model. The model views these determinants as prerequisites to any health improvements, and therefore proponents of the social model often argue that adoption of this model reduces social inequity and empowers individuals and communities (WHO, 2005). The increased use of stakeholders in the social model also encourages collaboration among various organizations (public, private, for-profit, non-profit) to influence social and environmental determinants to positively affect health statuses for individuals and communities.

Social model proponents typically support increased health education, healthy lifestyle promotion, and government support for community programs such as smoking cessation support and subsidized immunization programs. They also encourage individuals to take more control over their lifestyle choices that promote health and/or manage disease (such as diabetes and heart disease). The community approach makes the social model a popular construct for those in the public health field since it encourages the prevention of disease and promotion of health, which is a more cost-effective strategy than simply intervention based strategies.

Critics of the social model point out that health education is not always equally effective throughout the population, possibly leaving persons with lower levels health literacy behind. In addition, many attempts at lifestyle changes are not successful due to lack of motivation, inconvenience, or other reasons. Finally, the longitudinal efficacy of health promotions encouraging lifestyle changes is difficult to quantify (Fig. 3.2).

## 3.8 Spirituality and Health

Concepts of spirituality have existed in nursing and medical literature for at least 30 years (Purdy & Dupey, 2005), but are rarely conceptualized extensively in explicit terms (Anandarajah, 2008; McSherry, Cash, & Ross, 2004; Oldnall, 1995). Considerable evidence suggests that spirituality can play a useful supporting role in whole-person patient care, causing some researchers to propose adding spirituality to the biopsychosocial model, creating a biopsychosocial-spiritual model for health (Anandarajah, 2008). The World Health Organization and the American Association of American Medical Colleges have added spirituality to their medical education programs and recommendations (Anandarajah, 2008; Anderson et al., 1999; WHOQOL SRPB Group, 2006)

Spirituality has gained even further health research significant as complementary, alternative, and integrative medicines have gained popularity and their practitioners have promoted more holistic health models. These approaches to wellness maintain that "treatment is not just fixing what is broken; it is nurturing what is best" (Seligman & Csikszentmihaly, 2000, p. 9).

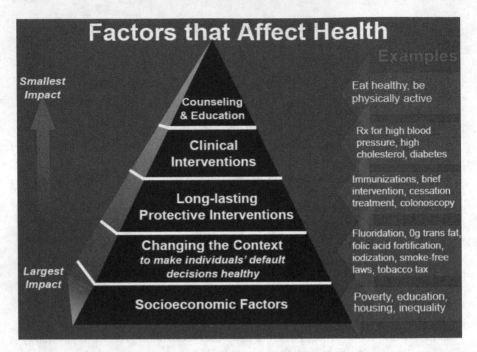

**Fig. 3.2** Factors that affect health. Public Domain. *Source*: CDC. Retrieved from http://www.cdc. gov/about/grand-rounds/archives/2010/download/GR-021810.pdf;     http://lulab.be.washington. edu/omeka/exhibits/show/lake2bay--public-health/item/1010

Attributes frequently associated with spirituality in the academic literature include terms such as *meaning, value, transcendence, connecting,* and *becoming* (Martsolf & Mickley, 1998). While spiritual models may not be well suited for all integrative care providers or all patients, some may be very useful in the counseling setting of an integrated health practice for those who find a focus on spirituality an appealing aspect of their health and wellness. Notably, spiritual models have been more frequently applied by counselors and therapists working with addiction recovery, grief, and depression (Martsolf & Mickley, 1998; Purdy & Dupey, 2005).

## 3.9   The Biopsychosocial-Spiritual Model

Very similar to the biopsychosocial model of health discussed in Chap. 1, the biopsychosocial-spiritual model of health adds a spiritual dimension to the conceptual framework (Wachholtz, Pearce, & Koenig, 2007). The spiritual component typically assesses a person's self assessed ability to discover meaning, connectedness, belonging, and faith.

## 3.10 The 3H Dimensions of Spirituality Model

The 3H Dimensions of Spirituality attempts to explain the various dimensions of spirituality. It is named after the dimensions of *Head*, *Heart*, and *Hands*, and was developed from a qualitative study which solicited responses from medical students and professionals, and a literature review. Within the *Head* dimension are beliefs, values, ideals, meaning, purpose, truth, wisdom, and faith (Anandarajah, 2008, p. 450). *Heart* incorporates connection, love, faith, inner peace, compassion, forgiveness and transcendence (Anandarajah, 2008, p. 450). *Hands* incorporates religious community, duties and behaviors, rituals and life and medical choices (Anandarajah, 2008, p. 450).

## 3.11 The Wheel of Wellness Model

The Wheel of Wellness is a "holistic, multidisciplinary model of wellness and prevention over the life span that is grounded in psychological theories of growth and behavior" (Smith, Myers, & Hensley, 2002). The wheel is best thought of as a three dimensional model in which it exhibits a combination of components of wellness that are reciprocally connected and consistently changing over time (Smith et al., 2002) (Fig. 3.3).

## 3.12 The Holistic Flow Model of Spiritual Wellness

Purdy and Dupey (2005) developed the Holistic Flow Model of Spiritual Wellness ("Holistic Flow Model"), making spirituality the central element considered in evaluating the patient's wellness and satisfaction. Six themes emerge from the Holistic Flow Model that are commonly referenced in the literature. These include "a belief in an organizing force in the universe, connectedness, faith, movement toward compassion, the ability to make meaning of life, and the ability to make meaning of death" (Purdy & Dupey, 2005, p. 99). As spiritual growth occurs within these themes, improvement is seen in areas such as personal, mental, environmental and psychological health. The Holistic Flow Model identifies these as: companionship, beauty/religion, body (physical health), emotions, mind, and one's life's work (Purdy & Dupey, 2005).

While this model may not be well suited for most integrative care providers to apply, it may be very useful in the counseling setting of an integrated health practice for those who find a focus on spirituality an appealing aspect of their wellness. Notably, models such as this have been used with addiction recovery, grief, and depression (Martsolf & Mickley, 1998; Purdy & Dupey, 2005). The levels of spiritual health within each of the three dimensions can be evaluated individually and the balance (or imbalance) between them can be measured (Fig. 3.4).

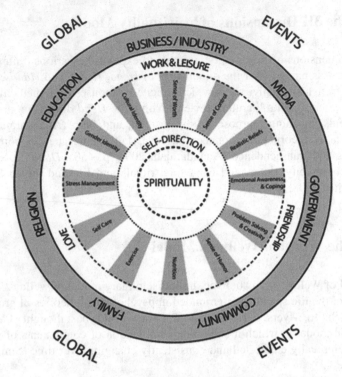

**Fig. 3.3** The wheel of wellness. *Source*: Witmer, Sweeney, Myers, 1998. Copyright 1998 by Witmer et al., p. 10. (Permission requested from Dr. Myers, unable to find contact information from other authors; no response at this time)

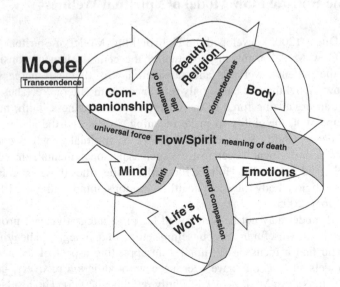

**Fig. 3.4** Components of holistic flow model of spiritual wellness. Reprinted with permission from © John Wiley and Sons. Originally published in Purdy, M., & Dupey, P. (2005). Holistic flow model of spiritual wellness. *Counseling and Values, 49*(2), 95–106

## 3.13 Questions for Discussion

1. After reading the various health models, which model or models are most representative with your values as a potential patient?

   a. What benefits do you see in the model(s) that are missing in other models?
   b. Are there any drawbacks to this model(s)?

2. How much of a role do you think social determinants of health play in the population?

   a. In your opinion is the importance of the role greater or smaller in different areas of the world? Or consistent throughout various populations?
   b. What role do you think social determinants have played in your own health or the health of a friend or family member?

3. Are particular health conditions better explained by particular models? If so, can you provide examples?

### 3.13.1 Case Study

You are a physician working in a primary care practice in urban Chicago. Your patient Raj is a 55 year old male who moved to the United States from India at age 20. Raj has suffered from high blood pressure for a number of years and is clinically obese. He is also an executive with a busy lifestyle who works many hours and travels frequently for work. He complains of having some trouble sleeping from time to time and occasional indigestion. He is in your office for his annual physical today. What questions do you ask Raj to uncover his health concerns and assess his health risks? What are your primary health concerns for this patient based on what you know? What recommendations do you plan to make?

## 3.14 Definitions

| | |
|---|---|
| Models of Health | Conceptual or theoretical frameworks providing a method to explain or predict health phenomenon and trends |
| Social Determinants of Health | the conditions in which people are born, live, work and age. The circumstances are shaped by the distribution of money, power, and resources at global, national, and local levels (WHO) |

| | |
|---|---|
| Biomedical Model | reductionist model in which disease is viewed as a result of biologic defect |
| Biopsychosocial Model | health and illness viewed as impacted by a variety of biological, psychological and social factors. The biopsychosocial model may be used as a philosophy and a clinical guideline. |
| Social Model | illness is attributable to social or psychological factors as well as other potential factors |
| Holistic Model | wellness viewed as impacted by six dimensions of wellness: physical, emotional, intellectual, spiritual, occupational, and social |
| Social Cognitive Theory | decisions are viewed as being made by weighing expectations and incentives |
| Health Belief Model | describes health behavior of individuals as a byproduct of health beliefs and typically used to predict an individual's health related behavior |
| Biopsychosocial-Spiritual Model | Similar to the Biopsychosocial model which considers biological, psychological, and social factors related to an individual's health, this model also considers spiritual factors related to health. Spiritual wellness may be indicated by the patient's sense of meaning and purpose, value, belonging, faith, and connection to others. |
| 3H Dimensions of Spirituality | Explains dimensions of spirituality as dimensions related to the head (beliefs, values, purpose, meaning, faith) heart (connection, love, faith, compassion, transcendence), and hands (duties, behaviors, rituals, decisions). |
| Wheel of Wellness Model | models of wellness and prevention over the life span which evolved from psychological theories of growth and behavior. |
| Holistic Flow Model of Spiritual Wellness | Model to evaluate spiritual wellness based on the themes of companionship, beauty/religion, body, emotions, mind, and life work. |

# References

Anandarajah, G. (2008). The 3 H and BMSEST models for spirituality in multicultural whole-person medicine. *The Annals of Family Medicine, 6*(5), 448–458.

Anderson, M. B., Cohen, J. J., Hallock, J. A., Kassebaum, D. G., Turnbull, J., & Whitcomb, M. E. (1999). Learning objectives for medical student education-guidelines for medical schools: Report I of the Medical School Objectives Project. *Academic Medicine, 74*(1), 13–18.

Borrell-Carrió, F., Suchman, A. L., & Epstein, R. M. (2004). The biopsychosocial model 25 years later: Principles, practice, and scientific inquiry. *Annals of Family Medicine, 2*(6), 576–582.

Engel, G. (1977). The need for a new medical model: A challenge for biomedicine. *Science, 196*, 129–136.

Gieck, J., & Olsen, S. (2007). Holistic Wellness as a means to developing a lifestyle approach to behavior among college students. *Journal of American College Health, 56*(1), 29–35.

Health Belief Model (Ch. 4). Hayden, J. A. (2013). *Introduction to health behavior theory*. Jones & Bartlett Publishers. Permissions granted.

Hochbaum, G. M. (1958). *Public participation in medical screening programs. A socio-psychological study*. Washington, DC: Public Health Service publication.

Johnson, S. B. (2012). Medicine's paradigm shift: An opportunity for psychology. *Monitor on Psychology, 43*(8), 5.

Martsolf, D. S., & Mickley, J. R. (1998). The concept of spirituality in nursing theories: Differing world-views and extent of focus. *Journal of Advanced Nursing, 27*(2), 294–303.

McSherry, W., Cash, K., & Ross, L. (2004). Meaning of spirituality: Implications for nursing practice. *Journal of Clinical Nursing, 13*(8), 934–941.

Oldnall, A. S. (1995). On the absence of spirituality in nursing theories and models. *Journal of Advanced Nursing, 21*, 417–418.

Purdy, M., & Dupey, P. (2005). Holistic flow model of spiritual wellness. *Counseling and Values, 49*(2), 95–106.

Rosenstock, I., Strecher, V., & Becker, M. (1988). Social learning theory and the health belief model. *Health Education Quarterly, 15*(2), 175–183.

Seligman, M., & Csikszentmihaly, M. (2000). Positive psychology: An introduction. *American Psychologist, 55*, 5–14.

Smith, S. L., Myers, J. E., & Hensley, L. G. (2002). Putting more life into life career courses: The benefits of a holistic wellness model. *Journal of College Counseling, 5*(1), 90–95.

Witmer, J. M., & Sweeney, T. J. (1992). Over the life span. *Journal of Counseling and Development, 71*.

Wachholtz, A. B., Pearce, M. J., & Koenig, H. (2007). Exploring the relationship between spirituality, coping, and pain. *Journal of Behavioral Medicine, 30*(4), 311–318.

WHOQOL SRPB Group. (2006). A cross-cultural study of spirituality, religion and personal beliefs as components of quality of life. *Social Science & Medicine, 62*(6), 1486–1497.

World Health Organization (WHO). (2005). *Commission on Social determinants of health*. Retrieved from http://www.who.int/social_determinants/resources/action_sd.pdf

World Health Organization (WHO). (1946). *Constitution of the World Health Organization*. Retrieved from http://www.who.int/governance/eb/who_constitution_en.pdf

# Chapter 4
# Health Care System Use and Disparities in Integrative Health Services

**Abstract** Most developed countries ensure access to health care for all citizens, and the United States is the major exception to this rule. When people do not have access to health care, they are typically left without preventative care, which is one of the central focuses of the integrative health model. This raises ethical questions related to access to health care in the U.S. Not only is going without preventive care putting people at risk, it is also creating additional costs for the health system in the long term. Health disparities related to race, ethnicity, and age will be discussed in this chapter, and concepts of health equity and health inequality will be introduced.

**Keywords** Health systems • Universal health care • Health equity • Health disparity • Bioethics • Access to care • Barriers to care • Cultural competence

## 4.1 Student Learning Objectives

After reading this chapter, you should be able to:

- Describe how the U.S. health care system measures up to other similarly situated countries in terms of efficiency, costs, access to care, and health outcomes
- Identify how the U.S. system is unique compared to other industrialized countries
- Better understand the ethical issues related to health care in the United States
- Understand the risks of direct payment health systems
- Differentiate between health equity and health equality

Most developed countries ensure access to health care for all citizens, and the United States is the major exception to this rule. This raises ethical questions related to access to health care in the U.S. In much of the developed world, healthcare is considered a human right. In these countries taxpayers typically finance care, and every resident has access to quality care and prescriptions medications. Exceptions include countries like Singapore which provide care for the poorest citizens and require others to deposit a percentage of their income into health savings accounts for their (or their family's) future use (Mullins-Owens, 2015). Many middle and low-income countries are not yet able to provide universal care. China has committed to ensuring universal access over the course of a number of years (China committed in 2009 to reaching universal coverage by 2020) (Mullins-Owens, 2015).

The U.S. health care system is by far the most expensive health care system in the world, but it is far from the best in term Dess of outcomes, access, patient satisfaction, efficiency, or equity (Davis, Stremikis, Squires, & Schoen, 2014). The Commonwealth Fund rated the U.S. health care system dead last among 11 wealthy industrialized countries (Davis et al., 2014). The data was based on information on health outcomes from the Commonwealth Fund, and data from the World Health Organization, and Organization for Economic Cooperation and Development, as well as patient and physician surveys (Davis et al., 2014).

## 4.2  Health System Ethics

Ethics in health care delivery is an issue of great concern in the United States. This is at least in part due to the fact that unlike most wealthy countries, the U.S. does not have a universal health care system that ensures all citizens access to quality care when they need it. This creates the need to examine how many people are largely left out of the health care system, and how that impacts their care, and public health at large. Many of those without insurance go without preventative health care, one of the cornerstones of the integrative health model. This creates more than just an ethical dilemma since the cost of treating most diseases far outweighs the cost of prevention or even early detection and treatment. In order to address these issues, we often look at determining the levels of health equity and the areas of health inequality.

Health inequity is sometimes confused with health inequality, however, the terms should be carefully distinguished. Health inequality refers to measurable differences and variations in the health achievements of individuals or populations (Kawachi, Subramanian, & Almeida-Filho, 2002). Health inequity is related to negative health outcomes which result from conditions which are unfair, unjust, and unnecessary (Mullins-Owens, 2015; Whitehead, 1992). Health equity is typically promoted and defended in ethical terms.

Most arguments for greater health equity are based upon on or more of the following three premises:

1. Health care is a right;
2. The resources for allocating health care are finite; and
3. Health policy should be concerned with the design of "just" mechanisms for allocating scarce health care resources (Aday & Andersen, 1981).

## 4.3   Universal Health Care

Planning a course towards universal health care coverage requires countries to first take stock of their current situation. Is there sufficient political and community commitment to achieving and maintaining universal health coverage? This question will mean different things in different contexts but will draw out the prevailing attitudes to social solidarity and self-reliance. A degree of social solidarity is required to develop universal health coverage, given that any effective system of financial protection for the whole population relies on the readiness of the rich to subsidize the poor, and the healthy to subsidize the sick. Recent research suggests that most, if not all, societies do have a concept of social solidarity when it comes to access to health services and health-care costs, although the nature and extent of these feelings varies across settings.

Since different societies have differing ethical notions of social justice, there is a vast difference in the amount of economic inequity is accepted in the health care context. Policy-makers then need to decide what proportion of costs will come from pooled funds in the longer run, and how to balance the inevitable tradeoffs in their use—tradeoffs between the proportion of the population, services and costs that can be covered. For those countries focused on maintaining their hard won gains, continual monitoring and adaptation will be crucial in the face of rapidly developing technologies and changing age structures and disease patterns (WHO, 2010). It is only when the reliance on direct payments falls to less than 15–20 % of total health expenditures that the incidence of financial catastrophe routinely falls to negligible levels (WHO, 2010).

## 4.4   Disparities in Care

Racial and ethnic disparities in healthcare manifest in a wide variety of manners. These disparities persist even after taking into account potential social and economic factors. The disparities are further complicated when culture or language is an additional contributing factor.

Disparities include higher rates of disease, disability, and premature death. There are also many quality gaps such as:

- Longer waits for care related to injury and illness;
- Less likelihood of receiving preventative care;
- Less likelihood of receiving prenatal care; and
- Poorly managed chronic diseases

(AHRQ National Healthcare Disparities Report, 2008). An Institute of Medicine report identified over 175 studies showing racial/ethnic disparities in the diagnosis and treatment of various conditions, even when analyses were controlled for:

- socioeconomic status
- insurance status
- site of care
- stage of disease
- co-morbidity
- Age (Institute of Medicine, 2002).

A "culturally competent" health care system has been defined as one that acknowledges and incorporates—at all levels—the importance of culture, assessment of cross-cultural relations, vigilance toward the dynamics that result from cultural differences, expansion of cultural knowledge, and adaption of services to meet culturally unique needs (Betancourt, Green, Carrillo, & Ananeh-Firempong, 2003). Culturally competent health systems are built on an awareness of the integration and interaction of health beliefs and behaviors, disease prevalence and incidence, and treatment outcomes for different patient populations (Betancourt et al., 2003). There is an inherent challenge in attempting to disentangle "social factors" from "cultural factors" (Betancourt et al., 2003). In *Defining Cultural Competence*, the authors' present three primary areas of care where sociocultural barriers occur that contribute to racial and ethnic disparities: organizational, structural and clinical barriers (Betancourt et al., 2003).

Organization barriers often result from issues in hiring and staffing. Health care leadership does not adequately represent the minority populations in the communities it serves. In terms of medical faculty, African Americans, Latinos, and Native Americans combined only make up 9 % of the full time faculty (Viets, Baca, Verney, Venner, Parker, & Wallerstein, 2009). These groups are even less represented at higher levels of management, accounting for less than 2 % (Betancourt et al., 2003; Viets et al., 2009). These same minorities make up 28 % of the national population. This impact of this disparity is especially significant since greater numbers of minority physicians ultimately choose to provide a greater percentage of care to poor and minority patients (Betancourt et al., 2003).

Structural barriers "arise when patients are faced with the challenge of obtaining health care from systems that are complex, underfunded, bureaucratic, or archaic in design" (Betancourt et al., 2003, p. 296). These structural barriers frequently occur when linguistic barriers keep patients from effectively communicating with providers. Minorities also report being more highly impacted by long wait times and difficulty obtaining specialty care (Betancourt et al., 2003).

Clinical barriers involve provider-patient interactions and may be unfairly influenced by differences in beliefs, attitudes, perceptions, and trust (in both care providers and in the medical system) (Betancourt et al., 2003). These barriers can be addressed by increasing cultural competence of health providers through training programs and by attempting to have a health provider and medical education workforce that is more reflective of the racial and ethnic diversity in the populations they serve (Betancourt et al., 2003).

It is estimated that many disparities are not even recognized by quality improvement measures. These may include reverse targeting by HMO; benchmarks for immunizations improperly obtained by immunizing large numbers of children at the lowest risk and fewer at high risk; and impacts of socioeconomic factors and race/ethnic makeup of members on plan performance (Fiscella, Franks, Gold, & Clancy, 2000, at p. 2580). Fiscella et al. (2000) provide five principles for addressing health disparities:

1. Disparities must be recognized as a significant quality problem
2. Collection of relevant and reliable data are needed to address disparities
3. Performance measures must be stratified by socioeconomic position, race, and ethnicity;
4. Population wide performance measures should be adjusted for socioeconomic position, race, and ethnicity; and
5. Approach to disparities should account for the relationships between both socioeconomic position and race/ethnicity and morbidity (Fiscella et al., 2000, pp. 2580–2581)

Older Americans also suffer many health disparities despite many being covered by Medicare. Ninety percent of older Americans never receive routine screening tests for bone density, colon or prostate cancer, or glaucoma (Currey, 2008). Sixty percent of older adults don't receive routine preventive health services, and 35 % of doctors continue to believe, despite ample evidence to the contrary, that elevated blood pressure is a "normal" part of aging (Currey, 2008).

Medical research mirrors this discrimination. Breast cancer is a disease that affects women over the age of 65 in more than one half of occurrences, yet in clinical trials evaluating new drugs for treating breast cancer, less than 10 % of participants are 65 or older (Currey, 2008). Indeed, clinical trials in general exclude or under-recruit older people as study participants (Currey, 2008). In a 2003 presentation at the American Thoracic Society, E. Wesley Ely, MD, MPH, of Vanderbilt University School of Medicine, noted that people 65 or older account for more than half of all intensive care unit (ICU) days nationwide, and people 75 or older account for seven times more ICU days than those under 65 (Currey, 2008). Despite this, further research done by Ely uncovered clear evidence of age bias in ICUs. While older ICU patients generally require more interventions and resources, "Older patients actually receive less aggressive care than do younger patients," he reported, noting that the use of mechanical ventilation in the ICU sharply decreases in patients 70 or older (Currey, 2008, p. 16). Meanwhile, lengthy data and compelling evidence confirms that preventive care addresses many elder healthcare deficiencies (Currey, 2008).

## 4.5  Direct Payments

The annual expenditure on health services in the U.S. is approximately $2.9 trillion (National Center for Health Statistics, 2015). The U.S. system is primarily privately financed by direct payers (patients), insurance companies and Health Management Organizations, and the government (largely in the form of Medicare, Medicaid, and

Tricare) (Lovett-Scott & Prather, 2012). Direct payments are those that occur at the point of delivery, most often by the uninsured. Frequently, direct payments discourage people from seeking care. Approximately 1.3 billion people globally are estimated to not have access to care when needed because they lack funds for direct payment at the point of service (WHO, 2010). Financial catastrophes may occur for about 150 million patients per year globally (WHO, 2010).

Financial catastrophe is most common in countries that rely on direct payment, although it is also sometimes a result of the high cost of care (such as in the United States) even for patients who possess insurance coverage (Lovett-Scott & Prather, 2012). In addition to being a cause for concern for the insured, it is an even greater cause for concern for those in the U.S. system without health insurance. Often this is seen among the working poor, who may not qualify for Medicaid programs that offer free care, and yet cannot afford to purchase their own insurance through the health exchanges.

## 4.6  Questions for Discussion

This chapter discussed organizational, structural and clinical barriers to care…

1. Which of the three types of barriers do you think presents the largest obstacle to care?
2. Which of the three types of barriers do you think would be the easiest to address? How would you propose addressing it?
3. Which of the three types of barriers do you think would be the hardest to address? How would you propose addressing it?

### 4.6.1  Case Study #1

You recently accepted a position as the executive director of a large urban hospital. The board of directors made it clear when they recruited you for the position that the person you are replacing was criticized for not adequately addressing health disparities in patient care. The hospital is located in a city that has a very large and diverse population, including immigrants from many different areas of the world. What will be your first steps as the executive director to ensure that these disparities are addressed? What support will you need from the board of directors, staff, health care providers and the community?

### 4.6.2  Case Study #2

You were just appointed as the senior health policy advisor to the President of the United States. He informs you he plans to push for major health reforms during his administration. He asks you to take a look at the data and provide suggestions to help form a plan to increase access to affordable quality health care for all citizens, and to increase affordable access to integrative health services in particular.

a. What stakeholders will you solicit to work with you on developing this plan and why?
b. What are your primary goals or concerns about increasing access to health care?
c. What are your primary goals or concerns about increasing access to integrative health services?
d. What will you consider in trying to keep costs down for implementing this plan?

## 4.7   Definitions

| | |
|---|---|
| Health Inequity | the disparities in care within a population which are unfair, unjust, and unnecessary |
| Direct payments | occur at the point of delivery. Frequently, direct payments discourage people from seeking care |
| Universal health care | health care system which ensures access to health care for all citizens along with financial protection against catastrophic expenses |
| Culturally competent care | Care that includes acknowledgment and incorporation of culture, assessment of cross-cultural relations, vigilance toward the dynamics that result from cultural differences, expansion of cultural knowledge, and adaption of services to meet culturally unique needs. |

## References

Aday, L. A., & Andersen, R. M. (1981). Equity of access to medical care: A conceptual and empirical overview. *Medical Care, 19*(12), 4–27.

Agency for Healthcare Research and Quality (AHRQ). (2008). *National Healthcare Disparities Report*. Retrieved from http://www.ahrq.gov/sites/default/files/wysiwyg/research/findings/nhqrdr/nhdr08/nhdr08.pdf

Betancourt, J. R., Green, A. R., Carrillo, J. E., & Ananeh-Firempong, O., II. (2003). Defining cultural competence: A practical framework for addressing racial/ethnic disparities in health and health care. *Public Health Reports, 118*(4), 293.

Currey, R. (2008). Ageism in healthcare: Time for a change. *Aging Well. 1*(1), 16. Retrieved from http://www.todaysgeriatricmedicine.com/archive/winter08p16.shtml

Davis, K., Stremikis, K., Squires, D., & Schoen, C. (2014, June). *Mirror, Mirror on the wall: How the performance of the U.S. health care system compares internationally.* 2014 Update. The Commonwealth Fund.

Fiscella, K., Franks, P., Gold, M. R., & Clancy, C. M. (2000). Inequality in quality: Addressing socioeconomic, racial, and ethnic disparities in health care. *Journal of the American Medical Association, 283*(19), 2579–2584.

Institute of Medicine. (2002). *Unequal treatment: Confronting racial and ethnic disparities in health care*. Washington, DC: National Academies Press.

Kawachi, I., Subramanian, S. V., & Almeida-Filho, N. (2002). A glossary for health inequalities. *Journal of Epidemiology and Community Health, 56*(9), 647–652.

Lovett-Scott, M., & Prather, F. (2012). *Global health systems: Comparing strategies for delivering health services*. Burlington, MA: Jones & Bartlett.

Mullins-Owens, H. (2015). Inequities in Chinese health services. *SAGE Open, 5*(1), 2158244015575187.

National Center for Health Statistics. (2015). *Health, United States, 2014: With special feature on adults aged 55–64*. Hyattsville, MD. Retrieved from http://www.cdc.gov/nchs/data/hus/hus14.pdf#102

Viets, V. L, Baca C., Verney, S. P., Venner, K., Parker, T., & Wallerstein, N. (2009). Reducing health disparities through a culturally centered mentorship program for minority faculty: The Southwest Addictions Research Group (SARG) experience. *Academic Medicine 84*, 8: 1118.

Whitehead, M. (1992). The concepts and principles of equity and health. *International Journal of Health Services, 22*(3), 429–445.

World Health Organization (WHO). (2010). *Health systems financing: The path to universal coverage*. Retrieved from: http://www.who.int/whr/2010/en/.

# Chapter 5
# Scope of Integrative Health Practice

**Abstract**  This chapter will provide an overview of integrative health practices in the clinical setting. It will include a discussion of some of the challenges of CAM and integrative health providers such as obtaining provider reimbursements and basic legal concerns. Concerns regarding patient safety and a lack of disclosure of CAM use to other health providers will be examined. In addition, the potential legal concerns for health providers who provide or refer patients to complementary care therapies will be considered. Finally, provider-patient communication and the concept of patient engagement will be discussed

**Keywords**  Integrative health • Provider reimbursement • Medical billing • Health providers • Health communication • Provider-patient communication • Training and education • Patient engagement • Provider referrals

## 5.1  Student Learning Objectives

After reading this chapter, you should be able to:

- Understand the challenges of obtaining provider reimbursement for integrative medical services
- Identify the concerns raised by the lack of consistent education and training among complementary care providers
- Explain the trend and practice of concierge physicians
- Identify the risks of patient nondisclosure of CAM use to their allopathic health care providers
- Describe the two types of health communication exchanges between health providers and patients

## 5.2   Provider Reimbursement

There are numerous challenges to implementing the integrative medical model broadly into the existing health care system, including the methodology of provider reimbursement and medical billing practices. Most patients use health insurance plans to pay for care at the point of service. These insurance plans use billing codes to determine what amount they will compensate the provider for services rendered. These billing codes make it difficult for providers to spend extra time with patients, to counsel them on health optimization, or to even get to know patient's family history and lifestyle well enough to anticipate potential issues that have not yet arisen (Kepner, 2003). If no billing code exists for the care provided, or in many cases for the time required to have a meaningful discussion about lifestyle changes, disease management, and self-care options, then providers cannot get paid for that time. When billing codes only provide very short time intervals for communicating with and educating patients, those conversations must be cut short. While billing codes may be a practical necessity, changes could be made to provide additional codes for more flexibility and autonomy for providers seeking to provide education and improve communication with patients.

Currently, many integrative care physicians find it too difficult to be paid under the traditional health model where billing codes do not exist for many or all of the services and education they provide patients. Some of these integrative physicians now choose to practice in boutique medical groups that offer subscription-based services to patients who pay for the prescriptions out-of-pocket, taking insurance billing challenges out of the equation (Horowitz, 2013). The cost of the subscription to medical services is the mechanism that allows physicians and other care providers to spend more time getting to know each patient, spend more time discussing preventative care, and often includes treatments from complementary practitioners that work within the medical group (Horowitz, 2013). These practices work on the direct-pay model discussed in the last chapter, making them more popular with higher income patients who can afford the cost without relying on insurance to share the cost burden.

Estimates show there were 4400 concierge physicians in 2012, a 30 % increase from the previous year (Horowitz, 2013). Data suggests these nontraditional boutique practices are becoming more prevalent over time (Horowitz, 2013). According to a 2012 Medscape study, 6 % of family doctors, 4 % of internists, and 2 % of pediatricians were in concierge or cash-only practices (Fuscaido, 2014). The following year, the 2013 Medscape report showed that 7 % of family doctors, 7 % of internists and 4 % of pediatricians were concierge or cash-based practices (Fuscaido, 2014). Concierge physicians typically earn roughly the same salary as physicians in traditional primary care practices, yet report greater job satisfaction (Horowitz, 2013). Patient care also improves, according to Horowitz (2013), who sites a 2012 study my a concierge practice which found a 79 % reduction in hospital admissions for Medicare patients; and a 72 % decrease for those with commercial insurance when compared with traditional primary care patients (Horowitz, 2013). While that may be a workable option for some patients, many who could really benefit from this type of medical care are left out due to inability to pay (Kepner, 2003). In addition, the new tax penalties contained in the Affordable Care Act (2010) for not obtaining medical insurance may make concierge practices less attractive than it may have been if they planned to forgo traditional coverage to offset their medical costs at a cash-only practice (Fuscaido, 2014).

## 5.3   Lack of Education About CAM in the Medical Community

With CAM usage increasing, some allopathic providers are seeking additional knowledge and training to share with their patients (Frenkel & Arye, 2001). Many medical schools are now offering courses in CAM, but the content is not consistent (Frenkel & Arye, 2001). In a literature review based on a survey of medical schools, Frenkel and Ben Arye (2001) report that medical students are often enthusiastic about learning about CAM, and education on the topic has been well received. The recommendation for a more consistent educational approach included the suggestion that a curriculum include knowledge of common CAM therapies that could be integrated into the conventional system in a useful, safe, and cost-effective way (Frenkel & Arye, 2001).

   Frenkel and Borkan (2003) worked with eight focus groups of specialists and trainees in two countries collaboratively developed a decision tree to assist primary care physicians in making recommendations to patients about CAM use. Frenkel and Borkan (2003) used literature review, key informant interviews, continuing education, focus groups, and field-testing to create an iterative model based on the varied feedback. The physicians are responsible under this model for evaluating the appropriateness of CAM modalities, and to monitor patient outcomes through ongoing contact (Frenkel & Borkan, 2003). While it may be difficult for primary care physicians to take the time to use this model with all patients, this could be very useful for specialty providers who work with patients who have serious or chronic conditions (e.g., cancer diagnosis, diabetes, osteoporosis, or heart disease). These specialty providers are likely to encounter the patients most likely to be CAM users, and most likely to be at risk of a dangerous interaction between CAM therapies and allopathic therapies. This is especially true for elderly patients with serious and/or chronic conditions, as they are most likely to be under care of both allopathic and CAM providers for the same medical condition (Adler, 2003).

## 5.4   Provider-to-Provider Collaboration in Patient Care

Provider-to-provider communication is a challenging part of the modern medical practice in which primary care physicians often have a handoff to other providers in different disciplines. Much of the research in this area focuses on communication in terms of patient safety (Koenig, Maguen, Daley, Cohen, & Seal, 2013). It is also thought that collaborative efforts between allopathic and CAM providers have many benefits including improved patient outcomes. Benefits include increased cultural sensitivity between allopathic and CAM providers, and better faculty development for medical teaching institutions involved (Nedrow et al., 2007). This data comes from a 2007 case study review of several projects funded by the National Institutes of Health National Center for Complementary and Alternative Medicine (Nedrow et al., 2007).

## 5.5  Lack of Consistent Training, Certification and Licensure Among Complementary Care Providers

One of the key challenges to further integration of complementary therapies into the mainstream medical community has been the lack of consistent training, certification and licensure. This lack of consistency in training may understandably make it difficult for physicians and others in the mainstream medical community to feel comfortable referring patients to certain therapies unless there is a practitioner they know and trust. In addition, it may also make insurance companies hesitant to reimburse patients for certain types of complementary health services in which training among providers is inconsistent.

In the U.S., homeopathy is generally not a licensed profession. Only three states offer a license (Arizona, Nevada, and Connecticut), and the license can only be obtained by medical doctors and doctors of osteopathy. Forty-six states have statutes regarding the professional regulation of dietitians or nutritionists, and the Academy of Nutrition and Dietetics is working to ensure similar provisions in the remaining four states (Eat Right Pro, n.d.).

Professional regulation of acupuncture varies from state to state as well. Forty-three states plus the District of Columbia require passage of the NCCAOM exam or certification to obtain a professional license (NCCAOM, Regulatory Affairs, n.d.). No license is required in Alabama, Kansas, North Dakota, Oklahoma, South Dakota, or Wyoming (Massage Therapy, n.d.). In Alabama, only a medical doctor osteopath, or chiropractor may practice acupuncture but there is no specific training required for doctors (Massage Therapy, 2015). Chiropractors are required to receive 100 h of training to perform acupuncture (Acufinder, n.d.).

Chiropractors are licensed and regulated in 18 countries, and is the largest and most tightly regulated of the CAM professions in the United States (Meeker & Haldeman, 2002). The average chiropractic curriculum consists of 4820 class and clinic hours, and most states also require continuing education (Meeker & Haldeman, 2002). Many chiropractors also offer other CAM services in their offices, such as massage, supplements, nutrition counseling, and lifestyle counseling (Meeker & Haldeman, 2002).

The required training, licensing and/or certification of massage therapists varies greatly from state to state. There are no state regulations regarding training, licensing or certification in Oklahoma, Kansas, Wyoming, Minnesota, or Vermont. State certification is voluntary in California, and required in Indiana and Virginia (Massage Therapy, n.d.). All other states require a license to be a massage therapist. Many states also require the MBLE or NCBTMB exam, although some states do not require any exam. Additional requirements in some states include background checks, CPR training, or graduation from an accredited school (Massage Therapy, n.d.).

Yoga Therapists in the United States are not licensed or regulated. The International Association of Yoga Therapists (IAYT) is a professional organization which offers voluntary registration to members who have either completed training through one of their accredited schools or been grandfathered into the organization (the school accreditation is very new). Yoga therapists tend to deal with patient populations which chronic or acute health conditions and therefore generally receive more anatomy training than other yoga teachers.

Yoga teachers have their own voluntary professional organization which works much the same way as the IAYT. If members graduate from an accredited yoga teacher school, they can become registered members with Yoga Alliance. Yoga Alliance offers continuing education and tracks the number of teaching hours of yoga teachers in its organization (all self-reported). These hours enable teachers to later qualify to upgrade their registration from Registered Yoga Teacher (RYT) to Experienced Registered Yoga Teacher (E-RYT) over time.

## 5.6 Patient Engagement

Evidence suggests that therapeutic activities are more effective when patients take active roles in their health care (Speedling & Rose, 1985). Patient Centered Care (PCC) is defined by the Institute of Medicine as "providing care that is respectful of and responsive to individual patient preferences, needs, and values, and ensuring that patient values guide all clinical decisions" (Institute of Medicine, 2001). Patient centered care may also be referred to as *patient engagement* or *patient activation* (Greene & Hibbard, 2012; Watson, 2014). The Chronic Care Model has been widely adopted and it calls for a reorganization of the health care system that would enable proactive providers to work in effective teams who will work with health literate, informed and motivated patients with the skill and confidence to make appropriate decisions as they manage their own health (Greene & Hibbard, 2012). The Center for Medicare and Medicaid Innovation review patient activation and engagement support in the scoring of new Pioneer Accountable Care Organizations (Greene & Hibbard, 2012). In a recent cross sectional study, Greene and Hibbard (2012) found that patient activation is strongly related to a variety of health-related outcomes.

The Kaplan et al. (1996) study researched the physician and practice characteristics associated with a practitioner's propensity to involve patients in diagnostic and treatment decisions, or participatory decision-making style. Data from 7730 patients from the practices of 300 physicians who practiced in general internal medicine, cardiology, primary care, and endocrinology were reviewed (Kaplan et al.,

1996). Patients in the study rated their physician's decision-making style, and additional data was retrieved from patient and physician screening visit questionnaires and clinician background questionnaires (Kaplan et al., 1996). The study used a patient-level regression model to calculate an expected participatory decision-making styles core for the physicians, then subtracted this score from the observed score (Kaplan et al., 1996).

Among the patients of the physicians who were least participatory, one third changed physicians in the next year (Kaplan et al., 1996). Only 15 % of patients in the highest quartile rating of physicians left their physicians in the same year (Kaplan et al., 1996). Higher volume practices had physicians with less participatory scores than the lower volume practices (Kaplan et al., 1996). Physicians with primary care or interview skill training were scored as more participatory than other physician on average (Kaplan et al., 1996).

One of the limitations of the Kaplan et al. (1996) study was that it measured decision-making style at only one point in time at one office visit. Also, the Medical Outcomes Study physician sample that was used for the study was not random and did not represent all potential physician variables (Kaplan et al., 1996). The study found that physicians who rated themselves as satisfied with their level of professional autonomy were rated more participatory (Kaplan et al., 1996).

## 5.7  Provider-Patient Communication

Provider-patient communication is a growing field of study in the health communication field. Medical schools have historically been criticized for a lack of focus on effective communication, although this improved in recent decades with the increasing professional focus on patient-centered care. Communication between providers and patients is of paramount importance in understanding the quality of care provided (Desjarlais-deKlerk & Wallace, 2013, p. 2). It also helps to encourage a productive and trustful provider-patient relationship (Desjarlais-deKlerk & Wallace, 2013). The quality of provider-patient communication is directly linked to patient satisfaction, adherence, and health outcomes (Betancourt, Green, Carrillo, & Ananeh-Firempong, 2003, p. 297).

Health communication between providers and patients involves (at minimum) two distinct types of exchanges: instrumental and socioemotional (Desjarlais-deKlerk & Wallace, 2013). Instrumental exchanges defined as "cure-oriented interactions," are information and data driven exchanges, such as those directly related to diagnosis, treatment, tests, and other medical care concerns (Desjarlais-deKlerk & Wallace, 2013, p. 3). Socioemotional communication is defined as "care-oriented interactions that have the primary goals of making the patient feel more comfortable, relieving patient anxiety and building a trusting relationship" (Desjarlais-deKlerk & Wallace, 2013, p. 3). Examples may include interactions that develop rapport between providers and patients, and providing social or emotional support. Early research suggests that increased socioemotional communication between providers and patients leads to better health outcomes (Desjarlais-deKlerk & Wallace, 2013; Ong, De Haes, Hoos, & Lammes, 1995; Roberts & Aruguete, 2000).

## 5.8   Lack of Disclosure by Patients to Physicians of Their CAM Use

Studies show that most patients do not discuss their use of CAM with their primary care providers. The communication gap between allopathic providers and patients about their CAM usage remains large (Wetzel, Kaptchuk, Haramati, & Eisenberg, 2003). This is a concern since CAM therapies may interfere with conventional treatments and could impact patient safety, well-being, and survival (Tasaki, Maskarinec, Shumay, Tatsumura, & Kakai, 2002). Furthermore, while some studies estimate that 77 % of patients do not discuss their CAM use with their physicians (Ben-Arye & Frenkel, 2008), research also shows they would like their primary care physicians input and/or referral to CAM modalities. Therefore, understanding reasons for nondisclosure are critical to improving practitioner-patient communication and patient satisfaction with their allopathic care providers, as well as ensuring patient safety (Tasaki et al., 2002).

In a 2004 literature review, 12 studies (conducted between 1993 and 2002) were reviewed related to disclosure of CAM therapies to other medical practitioners (Robinson & McGrail, 2004). Robinson and McGrail (2004) found nondisclosure rates ranging from 23 to 72 % in the various studies, and three common reasons for nondisclosure in the studies. These reasons were (1) concern about negative responses from the practitioners; (2) lack of inquiry from the practitioner; and (3) perception that allopathic practitioners have no knowledge of CAM therapies (Robinson & McGrail, 2004). Robinson and McGrail's (2004) literature review provides a good starting point for understanding the high rates of nondisclosure of CAM use, but the information gained from the analysis is only as strong as the data from the various studies that were reviewed. It also highlights the need for more research in this area and the need for better education about CAM therapies in medical schools (Kaplan et al., 1996).

Adler and Fosket (1999) conducted a study in which they conducted a series of four face-to-face interviews with subjects to determine why they do not disclose CAM use to allopathic practitioners. The subjects ranged from 35 to 74 years of age, with 52 % of subjects between 35 and 49 years of age (Adler & Fosket, 1999). After the series of interviews, analysis was done by obtaining transcripts of the interviews and using QSR NUD*IST software to conduct data analysis of the transcript text (Adler & Fosket, 1999).

Adler and Fosket (1999) found that 72 % of subjects used at least one CAM therapy at initial contact, and 65 % were using CAM 6 months later. Perhaps the most interesting finding of the study was that 94 % of the subjects reported discussing the details of their allopathic treatments with the CAM provider, but only 54 % disclosed CAM treatments to their allopathic practitioners (Adler & Fosket, 1999). One limitation of this study was that women between the ages of 50–59 were not included to facilitate cohort comparison even though they would have been more likely to use CAM than older women (according to other studies) (Adler & Fosket, 1999).

A 2002 study examined the communication barriers between physicians and cancer patients about CAM usage (Tasaki et al., 2002). The study recruited 143 cancer patients from among 1168 patients who completed a mail survey on CAM use (Tasaki et al., 2002). Thirty Non-CAM users were added based on similar geography. Semi-structured interviews were conducted to explore the patient experiences with CAM (Tasaki et al., 2002). Data was transcribed and analyzed using NU*DIST 4 (N4) software to find common words and phrases (Tasaki et al., 2002). Tasaki et al. (2002) found that the majority of patients reported negative experiences of discussing CAM with their physicians. Three themes about CAM use discussions emerged from the Tasaki et al. (2002) study:

1. Physicians' indifference or opposition to CAM therapies;
2. Physicians' emphasis on scientific evidence (or lack thereof);
3. Patients' anticipation of a negative response from their physicians

(Tasaki et al., 2002). One limitation of this study was that it did not collect data on physicians such as their gender or ethnicity (Tasaki et al., 2002). In addition, it did not attempt to counter patient perceptions with the physician perspective (Tasaki et al., 2002). Researchers recommended increasing medical education to include information about risks and benefits of CAM therapies to improve physician-patient communication (Tasaki et al., 2002). The article also recommends setting the tone for a discussion of CAM therapies with the physician by providing CAM materials in the practitioner's waiting room (Tasaki et al., 2002).

A recent Australian study of hospitalized patients found that CAM use among patients was over 90 %, but nearly 40 % of patients did not disclose the information to doctors or nurses (Shorofi & Arbon, 2010). The Shorofi and Arbon (2010) study was conducted using a five-page questionnaire containing both open and close-ended questions. The response rate for the questionnaire was 73.5 %, and participants ranged in age from 18 to 89, with a mean age of 47.8 (Shorofi & Arbon, 2010). Data analysis was done using the Statistical Package for Social Science (SPSS) software program (Shorofi & Arbon, 2010). The primary limitations of the study were that it relied on self-reporting, and that the study utilized a convenience sample due to difficulty estimating eligible participants (Shorofi & Arbon, 2010).

## 5.9  Legal Issues Regarding CAM When Provided by Physicians

Malpractice liability may occur when a given therapy does not meet the standard of care and the patient is injured. These standards come from a variety of sources such as academic literature, professional organizations and networks, and expert testimony. CAM therapies typically have standards of care that are more difficult to define that those in allopathic medicine. Often, there is less information available through professional and medical journals about these standards, and there may be fewer research studies to support what an evidenced based practice might look like. When evidence supports the safety and efficacy of a practice, a physician will probably not be liable for recommending and/or monitoring CAM therapies (Cohen & Eisenberg, 2002). If the evidence indicates serious risk or inefficacy, the physician will be more likely liable than in other instances. Examples of this might include injections of unapproved substances, or use of toxic herbs (Cohen & Eisenberg, 2002). However, the number of malpractice cases involving CAM therapies remains quite small (Gilmour, Harrison, Asadi, Cohen, & Vohra, 2011).

Physicians can minimize potential liability by taking some safety measure. Cohen and Eisenberg (2002) recommend the following precautions:

1. Determine the clinical risk level;
2. Document the literature supporting the therapeutic choice;
3. Provide adequate informed consent by engaging in a clear discussion of the risks and benefits of using the CAM therapy;
4. If feasible, obtain the patient's express written agreement to use the CAM treatment;
5. Continue to monitor the patient conventionally (pp. 599–600).

## 5.10 Legal Issues Regarding CAM Referrals by Physicians

It is possible that the physician owes a duty of care if he or she refers the patient to a particular provider of CAM, even though the physician did not provide that care directly (Cohen & Eisenberg, 2002; Gilmour et al., 2011). However, these types of law suits are very rare (Gilmour et al., 2011). Institutions can minimize liability by formulating a credentialing policy through which the institution verifies the qualifications of the CAM providers they work with and/or refer patients to (Gilmour et al., 2011). This should include proof of licensure or other credentialing and proof of practice insurance held by the CAM provider him or herself (Gilmour et al., 2011). The clinic or hospital working with the CAM provider should also ensure that their liability insurance covers CAM therapies (Gilmour et al., 2011).

## 5.11 Questions for Discussion

1. As a patient, what can you do to improve provider-patient communication in your health care interactions?
2. Are you concerned about the lack of standardized education, training, and/or certification of some types of CAM practitioners? Why or why not?

### 5.11.1  Case Study #1

You just accepted a position as dean of a large medical school. When hired, you were informed that the medical school is interested in adding education about CAM and integrative health to the curriculum, and you will be expected to take the lead on the project. You will need to decide what this new curriculum will include, and hire the necessary faculty to implement this change to the curriculum.

a. What would you like to be included in this CAM/Integrative health curriculum?
b. What qualities will you look for and what educational or professional background would you like to see in your new faculty?

### 5.11.2  Case Study #2

You have just graduated medical school with hopes of going into primary care practice and have two job interviews scheduled. One interview is for a position at a traditional mid-sized primary care clinic in your city. The other interview is for a position at a concierge medical clinic in the same city. Both jobs offer roughly equal anticipated pay and benefits.

a. What questions will you each employer at your interview?
b. If offered both positions, which are you inclined to prefer? Why?

## 5.12  Definitions

| | |
|---|---|
| Provider reimbursements | payments to health providers or health providers for medical treatments or hospital costs |
| Provider-Patient Communication | Communication between providers and patients. Positive provider-patient communication helps encourage productive and trustful provider-patient relationships, and is directly linked to patient satisfaction, adherence, and health outcomes |
| Patient Engagement | patient behaviors and actions taken to ensure optimal benefits from health services |
| Instrumental exchanges | Provider-patient communication focused on gathering, providing, or discussing information related to the medical treatment or care of the patient |

| Socioemotional communication | Provider-patient communication that focuses on building rapport, encouraging trust, or offering social or emotional support |
| Malpractice liability | May occur when a given therapy does not meet the standard of care and the patient is injured |

# References

Acufinder. (n.d.). Retrieved from https://www.acufinder.com/Acupuncture+Faqs#2

Adler, S. R. (2003). Relationships among older patients, CAM practitioners, and physicians: The advantages of qualitative inquiry. *Alternative Therapies in Health and Medicine, 9*(1), 104.

Adler, S. R., & Fosket, J. R. (1999). Disclosing complementary and alternative medicine use in the medical encounter: A qualitative study in women with breast cancer. *The Journal of Family Practice, 48*(6), 453–458.

Ben-Arye, E., & Frenkel, M. (2008). Referring to complementary and alternative medicine—A possible tool for implementation. *Complementary Therapies in Medicine, 16*(6), 325–330.

Betancourt, J. R., Green, A. R., Carrillo, J. E., & Ananeh-Firempong, O., II. (2003). Defining cultural competence: A practical framework for addressing racial/ethnic disparities in health and health care. *Public Health Reports, 118*(4), 293.

Cohen, M. H. (2002). Legal issues in complementary and integrative medicine: A guide for the clinician. *Medical Clinics of North America, 86*(1), 185–196.

Cohen, M. H., & Eisenberg, D. M. (2002). Potential physician malpractice liability associated with complementary and integrative medical therapies. *Annals of Internal Medicine, 136*(8), 596–603.

Desjarlais-deKlerk, K., & Wallace, J. E. (2013). Instrumental and socioemotional communications in doctor-patient interactions in urban and rural clinics. *BMC Health Services Research, 13*(1), 261.

Eat Right Pro. (n.d.) *Professional regulation of dietitians – An overview.* Retrieved from: http://www.eatrightpro.org/resource/advocacy/quality-health-care/consumer-protection-and-licensure/professional-regulation-of-dietitians-an-overview.

Emanuel, E. J., & Emanuel, L. L. (1992). Four models of the physician-patient relationship. *JAMA, 267*(16), 2221–2226.

Frenkel, M., & Arye, E. B. (2001). The growing need to teach about complementary and alternative medicine: Questions and challenges. *Academic Medicine, 76*(3), 251–254.

Frenkel, M. A., & Borkan, J. M. (2003). An approach for integrating complementary—Alternative medicine into primary care. *Family Practice, 20*(3), 324–332.

Fuscaido, D. (2014, April 15). Rise of direct-pay doctors: Good news for patients? *FoxBusiness.* Retrieved from http://www.foxbusiness.com/personal-finance/2014/04/15/rise-direct-pay-doctors-good-news-for-patients/

Gilmour, J., Harrison, C., Asadi, L., Cohen, M. H., & Vohra, S. (2011). Hospitals and complementary and alternative medicine: Managing responsibilities, risk, and potential liability. *Pediatrics, 128*(Suppl 4), S193–S199.

Greene, J., & Hibbard, J. H. (2012). Why does patient activation matter? An examination of the relationships between patient activation and health-related outcomes. *Journal of General Internal Medicine, 27*(5), 520–526.

Horowitz, A. S. (2013). Should you consider a concierge medicine practice? *The Profitable Practice.* Retrieved from Http://profitable-practice.softwareadvice.com/should-you-consider-concierge-medicine-0413/

Institute of Medicine. (2001). *Unequal treatment: Confronting racial and ethnic disparities in health care.* Washington, DC: National Academies Press.

Kaplan, S. H., Sullivan, L. S., Spetter, D., Dukes, K. A., Khan, A., & Greenfield, S. (1996). *Gender and patterns of physician-patient communication. Women's Health: The Commonwealth Fund Survey.* 76–95.

Kepner, J. (2003). Alternative billing codes and yoga: Practical issues and strategic considerations for determining "what is yoga therapy?" and "who is a yoga therapist?". *International Journal of Yoga Therapy, 13*(1), 93–99.

Koenig, C. J., Maguen, S., Daley, A., Cohen, G., & Seal, K. H. (2013). Passing the baton: A grounded practical theory of handoff communication between multidisciplinary providers in two Department of Veterans Affairs outpatient settings. *Journal of General Internal Medicine, 28*(1), 41–50.

Massage Therapy (2015). Retrived from: http://www.massagetherapy.com/careers/stateboards.php.

Massage Therapy. http://www.massagetherapy.com/careers/stateboards.php

Meeker, W. C., & Haldeman, S. (2002). Chiropractic: A profession at the crossroads of mainstream and alternative medicine. *Annals of Internal Medicine, 136*(3), 216–227.

NCCAOM. http://www.nccaom.org/regulatory-affairs/state-licensure-map

Nedrow, A. R., Heitkemper, M., Frenkel, M., Mann, D., Wayne, P., & Hughes, E. (2007). Collaborations between allopathic and complementary and alternative medicine health professionals: Four initiatives. *Academic Medicine, 82*(10), 962–966.

Ong, L. M., De Haes, J. C., Hoos, A. M., & Lammes, F. B. (1995). Doctor-patient communication: A review of the literature. *Social Science & Medicine, 40*(7), 903–918.

Roberts, C. A., & Aruguete, M. S. (2000). Task and socioemotional behaviors of physicians: A test of reciprocity and social interaction theories in analogue physician–patient encounters. *Social Science & Medicine, 50*(3), 309–315.

Robinson, A., & McGrail, M. R. (2004). Disclosure of CAM use to medical practitioners: A review of qualitative and quantitative studies. *Complement Ther Med, 12*(2–3), 90–98.

Shorofi, S. A., & Arbon, P. (2010). Complementary and alternative medicine (CAM) among hospitalised patients: An Australian study. *Complementary Therapies in Clinical Practice, 16*(2), 86–91.

Speedling, E. J., & Rose, D. N. (1985). Building an effective doctor-patient relationship: From patient satisfaction to patient participation. *Social Science & Medicine, 21*(2), 115–120.

Tasaki, K., Maskarinec, G., Shumay, D. M., Tatsumura, Y., & Kakai, H. (2002). Communication between physicians and cancer patients about complementary and alternative medicine: Exploring patients' perspectives. *Psycho-Oncology, 11*(3), 212–220.

Watson, Z. (2014). *What every provider needs to know about patient engagement. Technology Advice.* Retrieved from: http://technologyadvice.com/medical/blog/every-provider-needs-know-patient-engagement/.

Wetzel, M. S., Kaptchuk, T. J., Haramati, A., & Eisenberg, D. M. (2003). Complementary and alternative medical therapies: Implications for medical education. *Annals of Internal Medicine, 138*(3), 191–196.

# Chapter 6
# Public Health Workforce Implications

**Abstract** In recent years a trend of the singularly focused illness-based models of medicine are becoming replaced by more holistic wellness-based models. Integrative health and medicine in particular strive to ensure wellness-based approaches. This has led to increasing focus on patient-centered care strategies, greater focus on interdisciplinary and inter-specialty collaboration between health providers working together to treat patients, and increasing reliance on health care teams. The final chapter explores the challenges presented for future public health professionals. The lack of standard education for some modalities of CAM will be discussed, and evidence based practices will be presented. Provisions of Affordable Care Act supporting integrative medicine concepts will be explored. Finally, imperatives for the future of integrative health services will be explored.

**Keywords** Health models • Holistic models • Patient-centered care • Affordable Care Act (ACA) • Provider education • Evidence based practices • Systemic reviews • Provider-patient communication • Health communication • Provider collaboration

## 6.1 Student Learning Objectives

After reading this chapter, you should be able to:

- Describe the primary concepts of patient-centered care
- Identify primary challenges related to the lack of standard education for some modalities of complementary health therapies
- Understand the different types of health care teams and their levels of collaboration and communication
- Define evidence based practices
- Describe the role systemic reviews may play in promoting evidence based practices in the clinical setting
- Identify the ways in which the Affordable Care Act may promote some of the goals of integrative health

© Springer International Publishing Switzerland 2016                                                          49
H. Mullins-Owens, *Integrative Health Services*,
SpringerBriefs in Public Health, DOI 10.1007/978-3-319-29857-3_6

In recent years we have seen a trend of the singularly focused illness-based models of medicine being replaced by more wellness-based models. Integrative health and medicine in particular strive to ensure wellness-based approaches. This has led to increasing focus on patient-centered care strategies, greater focus on patient-provider communication, interdisciplinary and inter-specialty collaboration between health providers working together to treat patients, and increasing reliance on health care teams.

## 6.2   Patient-Centered Care

Patient-centered care (PCC) is defined by the Institute of Medicine as "providing care that is respectful of and responsive to patient preferences, needs, and values, and ensuring that patient values guide all clinical decisions" (Institute of Medicine, 2001). It may be referred to also as *patient engagement* or *patient activation* (Greene & Hibbard, 2012; Watson, 2014). Philosopher Plato was one of the first known proponents of what may be perceived as patient-centered medicine, writing that:

> A physician to slaves never gives his patient any account of his illness... the physician offers some orders gleaned from experience with an air of infallible knowledge... the free physician...who usually cares for free men, treats their diseases first by thoroughly discussing with the patient and his friends his ailment.

The widely adopted chronic care model calls for enabling proactive providers to work with health literate, informed and motivated patients with the skill and confidence to make appropriate decisions as they manage their own health (Greene & Hibbard, 2012).

> Informs and involves patients in medical decision making and self-management, coordinates and integrates medical care provides physical comfort and emotional support, understands the patients' concept of illness and their cultural beliefs, and understands and applies principles of disease prevention and behavioral change appropriate to diverse populations" (Maizes, Rakel, & Niemiec, 2009, pp. 277–278).

There are a variety of patient-centered approaches that can be described or measured. Some of the most commonly considered include approaches to communication style, therapeutic relationships, medical philosophy, quality of care, and shared decision-making between providers and patients.

According to Emanuel and Emanuel (1992), there are four models of provider-patient relationships which govern the aims of the relationship and decision-making style of communication: Paternalistic, Informative, Interpretive, and Deliberative. The paternalistic model is a hierarchical model that ensures the physician acts as the patient's guardian and is able to articulate and implement decisions for the good of the patient (Emanuel & Emanuel, 1992, p. 2221). The informative model makes the objective of the interaction one to provide the patient with all relevant information, with the patient making the decisions and the provider executing those decisions for the patient (Emanuel & Emanuel, 1992). The interpretive model attempts to "elucidate the patient's values and what he or she actually wants, and to help the patient select the available medical interventions that realize these values" (Emanuel & Emanuel, 1992, p. 2221). The fourth model, the deliberative model, attempts to

> Help the patient determine and choose the best health-related values that can be realized in the clinical situation…[with the physician to] delineate information on the patient's clinical situation and then help elucidate the types of values embodied in the available options. (Emanuel & Emanuel, 1992, p. 2222).

## 6.3 Interdisciplinary and Inter-specialty Collaboration

Not enough is known about the benefits and obstacles to collaboration between biomedical and complementary health providers. The National Institutes of Health National Center for Complementary and Integrative Health (NCCIH) funded 15 projects between 2000 and 2002 to educate allopathic medical and nursing learners in CAM literacy (Nedrow et al., 2007).

Four formed collaborative partnerships with nearby CAM academic institutions. Nedrow et al. (2007) examines these examples of institutional collaborations and their outcomes, and found that increasing collaboration can lead to better faculty development, future credentialing for complementary providers, improved patient outcomes, and increased cultural sensitivity between allopathic and complementary disciplines (Nedrow et al., 2007).

With increasing use of complementary care by patients, medical providers are seeking more information about complementary health practices to share with patients and integrate into patient care (Frenkel & Arye, 2001). Some medical schools are even providing CAM courses for biomedical practitioners, but it is not certain what type of content these courses should contain (Frenkel & Arye, 2001). In a 1998 survey of US. Medical schools, 123 courses in CAM were being offered (Frenkel & Arye, 2001). The Frenkel and Arye (2001) literature review showed a significant interest among medical students in learning more about CAM, and recommended more consistent educational approach with more standardized content.

## 6.4 Health Care Teams

The term 'health care team' has been define and interpreted in a wide variety of manners (Boon, Verhoef, O'Hara, & Findlay, 2004). Boon et al. (2004) provides a formalized description of the seven common distinctions in provider communication that are sometimes described under the umbrella term of 'integrative care' (p. 3). These can be considered as a continuum (see representation below) with the far left model characterized by individual providers working independently in a common setting, and at the far right an interdisciplinary collaboration between conventional and complementary/alternative health providers offering a "seamless continuum of

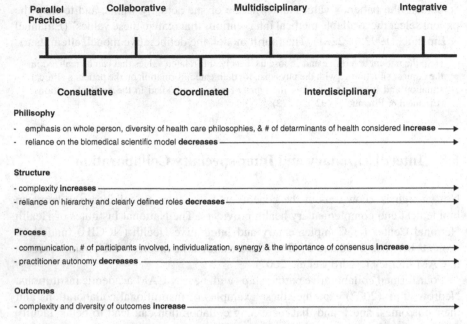

**Fig. 6.1** A continuum of health care practice models. Source: Boon et al. BMC Health Services Research 2004 4:15. Reprinted per BioMed's open access charter, provided here: https://www.biomedcentral.com/bmchealthservres/about/faq/reproduce

decision-making and patient-centered care and support....based on a specific set of core values that include the goals of treating the whole person...promoting health and wellness...[and] prevention of disease" (Boon et al., 2004, p. 3) (Fig. 6.1).

Mann, Gaylord, and Norton (2004) offer seven integrative health models describing the varying degrees of allopathic and complementary care integration. These are:

> the informed clinician; the informed networking clinician; the informed CAM-trained physician; multidisciplinary integrative group practice; interdisciplinary integrative group practice; hospital-based integration; and integrative medicine in an academic medical center (Adams, Hollenberg, Lui, & Broom, 2009, p. 793).

The Institute of Medicine (2001) has recommended changes to the health care system which emphasized strong relationships and communication among health care teams. Ideally, Wagner et al. (2001) recommends that teams treating patients with chronic disease follow a four-step process:

1. Elicit and review data concerning patients' perspectives and other critical information about the course and management of the condition(s);
2. Help patients to set goals and solve problems for improved self-management;
3. Apply clinical and behavioral interventions that prevent complications and optimize disease control and patient well-being; and
4. Ensure continuous follow-up (Wagner et al., 2001, p. 66).

There are multidisciplinary care teams in conventional medicine that offer examples of integrative care. These include:

(1) in geriatrics, developing multidisciplinary special care teams and end-of-life programs to optimize quality of life in hospice care (2) in chronic pain treatment programs, applying multidisciplinary care to improve individual self-efficacy and quality of life; (3) in psychiatry, blending social supports, psychotherapy, and medications as well as emphasizing the patient's responsibility for his or her own recovery; (4) in family medicine, valuing good physician-patient relationships and preventive interventions and (5) in behavioral medicine/health psychology, using behavioral interventions to foster self-care and self-efficacy in patients with diabetes or arthritis (Bell et al., 2002, pp. 133–140).

## 6.5   Benefits of Using CAM Practitioners to Foster Health Promotion and Disease Prevention

The data shows there would be many benefits to mainstream acceptance of an integrative medical model of care in the United States. While some innovative integrative medicine clinics are now bringing primary care physicians and CAM practitioners together and merging the practices, it is unknown how many of these clinics currently exist across the country (Ben-Arye & Frenkel, 2008). While some clinics such as this appear very successful (e.g., the Arizona Center for Integrative Medicine), many others struggle to find appropriate ways to bill patients for services not covered by most insurance companies, or for visits with physicians that may take much longer than the customary time for appropriate billing codes

In order to see integrative medicine succeed in the mainstream medical community, there are several unresolved issues to address. Primary among these is the need for patients to be encouraged to disclose information about their CAM use to allopathic providers. In order to effectively manage the care of their patients, allopathic practitioners need to receive education about CAM efficacy, risks, and benefits. Ideally, medical schools and nursing schools would develop standardized CAM education programs. Collaboration between CAM educational institutions and medical/nursing schools would be very helpful to the institutions and the health practitioners being educated there. This would allow greater understanding of different therapies, and possibly create better future dialogue between the future allopathic and CAM practitioners. Integration of the eight principles of integrative medicine need to be better implemented into allopathic institutions, particularly in primary care clinics. Finally, changes need to be made to the way medical insurance and medical billing are conducted. This would ideally involve more coverage for evidence based CAM therapies, and more time for practitioners to communicate and collaborate with patients and their other health providers.

## 6.6   Evidence-Based Practices

Evidence Based Practice (EBP) seeks to integrate clinical expertise; external scientific evidence; and client/patient/caregiver perspectives to provide high-quality services reflecting the interests, values, needs and choices of the individuals that are served (American Speech Language-Hearing Association, 1997–2016). SP began in the 1970s (Cochrane, 1972) and gained substantial popularity in health care in the 1990s. Even as EBP is now routinely encouraged by health care administrators for clinicians, administrators themselves have been slow to apply these practices.

EBP might be defined as the "access to, and application and integration of evidence to guide clinical decision making to provide best practice for the patient/client. Evidence based practice includes the integration of best available research, clinical expertise, and patient/client management, practice management, and health care policy decision making" (American Physical Therapy Association, 2010). In the 1970s when EBP was first introduced, there were may publications and very little way for physicians to effectively and efficiently navigate all of the data. However, with increased availability of desktop computers and the internet in the 1980s, applying EBP became a more realistic goal for clinicians. Evidence Based Practice Methodology has a five step process. Evidence Based Practice Methodology has a five-step process:

1. Ask a clinical question
2. Acquire information that may answer the question
3. Appraise the evidence for its quality and applicability
4. Apply the evidence in the care of the patient
5. Assess whether the application resulted in the expected outcomes (Dijkers, Murphy, & Krellman, 2012).

As with all models there are some inevitable gaps between research and practice. In terms of EBP, the research-practice gap often results in three categories of problems:

1. overuse of health interventions
2. underuse of health interventions; and
3. misuse of health interventions (See Table 1 Walshe & Rundall, 2001)

While it may seem obvious that evidence based practices should be applied to clinical situations, it is often a challenge. There is a tremendous volumes of research conducted annually, and some of the research that is conducted may have conflicting results. In addition, not all research studies are equally reliable or relevant, and determining the amount of weight to give to each study can be very time consuming. The complexity of large health care organizations and their processes mean that often substantial and significant changes to the processes are required to implement EBP. Also, organization leaders and clinicians may have very little time to keep themselves up-to-date on the most recent research in addition to their other job responsibilities (Walshe & Rundall, 2001).

Systemic reviews such as the Cochrane Reviews, provided by the Cochrane Collaboration, have helped to make finding useful and vetted information about evidence based practices much easier for busy health care managers and clinicians. Systemic reviews help to "identify, appraise and synthesize all the empirical evidence that meets pre-specified eligibility criteria to answer a given research question" (Cochrane Library, 1999–2015). These reviews look at the extensive clinical research often available on a health topic, measure potential bias and assess the possible accuracy of the data, and then organize that information in a way that is intended to best facilitate clinical decision making based on the available evidence. The Cochrane Database of Systemic Reviews is one of the most widely available and frequently utilized of these types of resources. Their library includes interventions reviews; (assessing benefits and dangers of interventions), diagnostic test accuracy reviews; methodology reviews (reviewing research methods and reporting); qualitative reviews (research not focused on efficacy); and prognosis reviews (providing information about likelihood and timing of disease progression and future outcomes) (Cochrane Library, 1999–2015). Most importantly, the information collected and published by the Cochrane Library is available online and free of charge in many low income countries,[1] so that more clinicians and public health workers around the world are better able to access evidence based evidence and provide the best informed health care possible (Fig. 6.2).

The Research-Practice Gap: Examples of Overuse, Underuse, and Misuse Drawn from Reviews by the NHS Centre for Reviews and Dissemination

| | |
|---|---|
| Overuse | • Prophylactic extractions of asymptomatic impacted third molars (wisdom teeth) |
| | • Screening for prostate cancer |
| | • Composite and other new materials used for dental fillings in place of traditional amalgam |
| | • Atypical antipsychotic drug treatments for schizophrenia |
| Underuse | • Drug treatment of essential hypertension in older people |
| | • Smoking cessation through nicotine replacement therapy |
| | • Compression therapy for venous leg ulcers |
| | • Cardiac rehabilition for people with heart disease |
| Misuse | • Pressure-relieving equipment in the prevention of pressure sores |
| | • Interventions to diagonose and treat gynecological cancers |
| | • Selection of hip prostheses in hip replacement surgery |
| | • Some preschool hearing, speech, language, and vision screening tests |

**Fig. 6.2** The research-practice gap. Source: Walshe, K., & Rundall, T. G. (2001). Evidence-based Management: From Theory to Practice in Health Care. *Milbank Quarterly, 79*(3), 431. Reprinted with permission. © 2001 Milbank Memorial Fund. Published by John Wiley & Sons

---

[1] Free or reduced cost access is available through HINARI, a collaboration between the World Health Organization and many medical and scholarly publishers. A list of countries qualifying for free access is available here: http://www.who.int/hinari/eligibility/en/.

## 6.7  Affordable Care Act

In March 2010, the Patient Protection and Affordable Care Act (ACA) was signed into law by President Obama. The ACA attempts to improve the quality of healthcare in America by eliminating lifetime limits on coverage; limiting annual limits on benefits; prohibiting denial of coverage or excessive fees for customers with pre-existing conditions; extending dependent coverage to age 26; capping non-medical expenditures by insurance companies from premium revenues; and prohibiting different insurance premiums based on gender, health status or salary. While many argue that the ACA did not go far enough to ensure greater access to health care, 16.9 million people who did not have insurance prior to the exchanges opening in September 2013 now have insurance coverage (Fox, 2015). It does not appear from the early data that the ACA has been able to cap health care spending in any meaningful way. The Act did require insurance plans to provide prescription drug coverage, but the consumer cost of the drugs varies among the four levels of coverage, and the preferred-drug lists vary (CVS Caremark, 2014).

The ACA's focus on preventive health care without copayments and coinsurance was a welcome change in the integrative health field. Fifteen preventive services and an annual wellness visit be provided to all adults without copays or coinsurance are provided to all adults under the ACA, and this coverage is available without meeting the plan's deductible. These preventive services include immunizations; screenings for depression, breast cancer, blood pressure, cholesterol, diabetes, HIV, HPV, cervical and colorectal cancer; diet and alcohol abuse counseling; and contraception. In addition, children are now entitled to 26 preventive services without copays or coinsurance, which include a variety of screenings and immunizations (U.S. Department of Health and Human Services, 2015). Although more data is needed across various screening populations, early data indicates that the ACA's removal of the out-of-pocket payment barriers encouraged more Americans to take advantage of the colorectal cancer and breast cancer screenings (Han, Yabroff, Guy, Zheng, & Jemal, 2015; Nelson, Weerasinghe, Wang, & Grunkemeier, 2015).

In addition to these screenings, the ACA provides incentives for employers to fund employee wellness programs. The return on investment is estimated to be at least $3 in medical costs for every $1 spent on such programs (Anderko et al., 2012). As the workforce continues to age and rates of chronic disease continue to rise, these savings could be even greater over the next decades. It is anticipated that depression and stress will two areas of particular focus in the workplace plans since they affect large numbers of employees and cause high losses in productivity (Anderko et al., 2012).

## 6.8   Future of Integrative Health Services

Increasing numbers of academic institutions are incorporating CAM into medical education, research, and clinical practice (Ventola, 2010, p. 466). The federal government has provided $22 million to help fund the inclusion of CAM into medical and nursing school curriculums (Ventola, 2010). The Federation of State Medical Boards (FSMB) has issued guidance on patient counseling about CAM therapies (FSMB, 2002). Many physicians acknowledge a need for more training. In one study of physicians and pharmacists, 76 % rated physicians as poorly informed about herbal supplements, and 46.6 % rated their personal knowledge as "very poor" or "quite poor" (ModernMedicine Network, 2010). Perhaps most surprising, 77.3 % stated they worried patients will take such supplements without disclosure to them, but only 12.9 % reported that they always inquired about patient use (ModernMedicine Network, 2010).

Studies show that pharmacists may be the ideal health professionals to educate consumers about health supplements since the majority of these are purchased in pharmacies (Ventola, 2010). Studies also show consumers are more comfortable asking pharmacists about these therapies as many are concerned that their primary care providers may disapprove (Ventola, 2010). Pharmacists, however, report that they are uncomfortable answering questions about these supplements because there is not enough data and research to make them feel knowledgeable in making recommendations (Ventola, 2010).

Ethical issues abound as the public must wait for mainstream health providers to become better informed about CAM therapies. A commitment to joint problem-solving between physicians, patients, and possibly pharmacists is needed to not only ensure quality patient care, but also to minimize potential risks to patients.

It is clear that patients see increasing value in integrative health services. Visits to CAM providers has exceeded visits to primary care physicians since the 1990s (Eisenberg et al., 1998). Consumers are also spending more on CAM visits annually than primary care visits (Eisenberg et al., 1998). It is even clearer that patients value these services since most are currently paying out of pocket for these services that are not covered by their medical insurance. Given that fact, it is likely that even more patients would choose integrative and CAM services if they were more economically accessible.

In addition to the ethical arguments for greater integration of allopathic and CAM practices in an integrative health model to ensure the best and safest care for patients, it there are clear economic incentives for medical providers willing to make this transition (Weil, 2000). Patients are showing that they value natural approaches to promoting their long term health and working towards their own optimal wellness. Increasingly, the patients themselves perceive their health to include more than just the absence of disease. It is imperative that the new public health workforce recognizes these shifts in consumer and patient perspectives on health and support them by providing appropriate and safe care.

## 6.9  Questions for Discussion

1. From the public health perspective, how important do you think it is to encourage participation in preventative health screenings? What more could be done to achieve this?
2. The Affordable Care Act ensures that insurance companies provide some preventative screenings without copays or coinsurance. Does this make it more likely that you will take advantage of these opportunities? Why or why not?
3. There is always a research-practice gap when trying to apply medical research into appropriate patient care. What are some ways we try to minimize this gap? Do you have additional suggestions to minimize the gap?

### 6.9.1  Case Study

As a hospital administrator you were asked to select three departmental care units within the hospital to receive additional education on patient centered care. How will you determine which care units might most benefit from this type of education, assuming that the level of patient care is currently similar between care units? Is there a particular emphasis that you think would be useful for the training session? How will you try to assess the impact of this training once it is completed?

## 6.10  Definitions

Affordable Care Act          Legislation signed by President Obama in 2010 which provides new rules for insurance companies and seeks to provide affordable insurance plans to more consumers, increase access to preventive care, encourage more wellness screenings, limit administrative spending by insurance companies, and ends lifetime coverage limits and arbitrary coverage termination.

Evidence Based Practices     application and integration of evidence to guide clinical decision making to provide best practice for the patient/client

Patient-Centered Care        care that is responsive to patient preferences and values, encouraging the patient to participate in clinical decision-making

Systemic Review              Looks at the extensive clinical research and then measures bias and assesses the accuracy of the data, publishing that information to facilitate efficient research on evidence based practices in the clinical setting

# References

Adams, J., Hollenberg, D., Lui, C. W., & Broom, A. (2009). Contextualizing integration: A critical social science approach to integrative health care. *Journal of Manipulative and Physiological Therapeutics, 32*(9), 792–798.

American Physical Therapy Association (APTA) Vision 2010. Retrieved from http://www.apta.org/vision2020/

Anderko, L., Roffenbender, J. S., Goetzel, R. Z., Millard, F., Wildenhaus, K., DeSantis, C., & Novelli, W. (2012). Peer reviewed: Promoting prevention through the affordable care act: Workplace wellness. *Preventing Chronic Disease, 9,* E175.

Bell, I. R., Caspi, O., Schwartz, G. E., Grant, K. L., Gaudet, T. W., Rychener, D., ... Weil, A. (2002). Integrative medicine and systemic outcomes research: Issues in the emergence of a new model for primary health care. *Archives of Internal Medicine, 162*(2), 133–140.

Ben-Arye, E., Frenkel, M., Klein, A., & Scharf, M. (2008). Attitudes toward integration of complementary and alternative medicine in primary care: Perspectives of patients, physicians and complementary practitioners. *Patient Education and Counseling, 70*(3), 395–402.

Ben-Arye, E., & Frenkel, M. (2008). Referring to complementary and alternative medicine—A possible tool for implementation. *Complementary Therapies in Medicine, 16*(6), 325–330.

Boon, H., Verhoef, M., O'Hara, D., & Findlay, B. (2004). From parallel practice to integrative health care: A conceptual framework. *BMC Health Services Research, 4*(1), 15.

Cochrane, A. L. (1972). *Effectiveness and efficiency: Random reflections on health services.* London: Nuffield Provincial Hospitals Trust.

Cochrane Collaboration. *Evidence based health care.* Retrieved from http://www.cochrane.org/about-us/evidence-based-health-care

Cochrane Library. (1999–2016). *About Cochrane reviews.* John Wiley & Sons, Inc. Retrieved from http://www.cochranelibrary.com/about/about-cochrane-systematic-reviews.html

CVS Caremark. (2014, March 17). What the Affordable Care Act means for prescription coverage. *Washington Post.* Retrieved from http://www.washingtonpost.com/sf/brand-connect/wp/2014/03/17/what-the-affordable-care-act-means-for-prescription-coverage/

Dijkers, M., Murphy, S., & Krellman, J. (2012). Evidence-based practice for rehabilitation professionals: Concepts and controversies. *Archives of Physical Medicine and Rehabilitation, 93*(8), S164–S176.

Emanuel, E. J., & Emanuel, L. L. (1992). Four models of the physician-patient relationship. *JAMA, 267*(16), 2221–2226.

Eisenberg, D. M., Davis, R. B., Ettner, S. L., Appel, S., Wilkey, S., Van Rompay, M., & Kessler, R. C. (1998). Trends in alternative medicine use in the United States, 1990–1997: Results of a follow-up national survey. *JAMA, 280*(18), 1569–1575.

Federation of State Medical Boards (FSMB). (2002, April). *Model guidelines for the use of complementary and alternative medicine in medical practice.*

Fox, M. (2015, May 6). Nearly 17 million Americans covered under Obamacare. *NBC News.* Retrieved from http://www.nbcnews.com/storyline/obamacare-deadline/4-p-embargo-nearly-17-million-americans-covered-under-obamacare-n354851

Frenkel, M., & Arye, E. B. (2001). The growing need to teach about complementary and alternative medicine: Questions and challenges. *Academic Medicine, 76*(3), 251–254.

Fuscaido, D. (2014, April 15). Rise of direct-pay doctors: Good news for patients? *FoxBusiness.* Retrieved from http://www.foxbusiness.com/personal-finance/2014/04/15/rise-direct-pay-doctors-good-news-for-patients/

Greene, J., & Hibbard, J. (2012). Why does patient activation matter? An examination of the relationships between patient activation and health-related outcomes. *Journal of General Internal Medicine, 27*(5), 520–526.

Han, X., Yabroff, K. R., Guy, G. P., Zheng, Z., & Jemal, A. (2015). Has recommended preventive service use increased after elimination of cost-sharing as part of the Affordable Care Act in the United States? *Preventive Medicine, 78,* 85–91.

Institute of Medicine. (2001). *Crossing the quality chasm: A new health system for the 21st century*. Retrieved from http://www.nap.edu/catalog/10027.html.

American Speech-Language-Hearing Association. (1997–2016). *Introduction to evidence-based practice: What it is (and what it isn't)*. Retrieved from http://www.asha.org/Research/EBP/Introduction-to-Evidence-Based-Practice/.

Maizes, V., Rakel, D., & Niemiec, C. (2009). Integrative medicine and patient-centered care. *Explore, 5*(5), 277–289.

Mann, D., Gaylord, S., & Norton, S. (2004). Moving toward integrative care: Rationales, models, and steps for conventional-care providers. *Complementary Health Practice Review, 9*(3), 155–172.

Most doctors not knowledgeable about herbals. ModernMedicine Network. (2010). Retrieved from http://www.modernmedicine.com/%5Bnode-source-domain-raw%5D/news/clinical/clinical-pharmacology/most-doctors-not-knowledgeable-about-he?page=full

Nedrow, A. R., Heitkemper, M., Frenkel, M., Mann, D., Wayne, P., & Hughes, E. (2007). Collaborations between allopathic and complementary and alternative medicine health professionals: Four initiatives. *Academic Medicine, 82*(10), 962–966.

Nelson, H. D., Weerasinghe, R., Wang, L., & Grunkemeier, G. (2015). Mammography screening in a large health system following the US Preventive Services Task Force Recommendations and the Affordable Care Act. *PLoS One, 10*(6), e0131903.

U.S. Department of Health and Human Services. (2015, September 22). Preventive services covered under the Affordable Care Act. Fact Sheets. Retrieved from http://www.hhs.gov/healthcare/facts/factsheets/2010/07/preventive-services-list.html

Ventola, C. L. (2010). Current issues regarding Complementary and Alternative Medicine (CAM) in the United States: Part 1: The widespread use of CAM and the need for better-informed health care professionals to provide patient counseling. *Pharmacy and Therapeutics, 35*(8), 461.

Wagner, E. H., Austin, B. T., Davis, C., Hindmarsh, M., Schaefer, J., & Bonomi, A. (2001). Improving chronic illness care: Translating evidence into action. *Health Affairs, 20*(6), 64–78.

Walshe, K., & Rundall, T. G. (2001). Evidence-based management: From theory to practice in health care. *Milbank Quarterly, 79*(3), 429–457.

Watson, Z. (2014). What every provider needs to know about patient engagement. *Technology Advice*. Retrieved from http://technologyadvice.com/medical/blog/every-provider-needs-know-patient-engagement/

Weil, A. (2000). The significance of integrative medicine for the future of medical education. *The American Journal of Medicine, 108*(5), 441–443.

# Index

**A**

ACA. *See* Affordable Care Act (ACA)
Acupuncture, 13
Affordable Care Act (ACA), 12
  definition, 58
  depression and stress, 56
  incentives, 56
  insurance plans, 56
  preventive services, 56
  quality of healthcare, 56
Allopathic medical care, 4
Allopathic Medicine, 7
Arizona Center for Integrative
    Medicine (ACIM), 3, 4

**B**

Biomedical model, 18, 26
Biopsychosocial model, 19, 26
Biopsychosocial-spiritual model, 22, 26

**C**

CAM. *See* Complementary and alternative
    medicine (CAM). *See*
Chiropractic care, 3
Chronic care model, 12, 15
Chronic conditions, 4, 8, 12
Chronic disease management, 12
Chronic diseases, 18
Complementary and alternative medicine
    (CAM), 1, 7, 43, 44
  disclosure of

  communication gap, providers
      and patients, 43
  data analysis, 44
  face-to-face interviews, 43
  limitation, 44
  nondisclosure rates, 43
  semi-structured interviews, 44
  lack of education, 39
  malpractice liability, 44, 45
  popularity, 2–3
  referrals, 45
Complementary and integrative approach to
    treatment, 13
Costs of Care, 13
Culturally competent care, 35

**D**

*Defining Cultural Competence*, 32
Direct payments, 34, 35
Disparities in care, 31–33

**E**

EBP. *See* Evidence based practice (EBP)
End of Life Care, 13–14
Ethics in health care delivery, 30
Evidence based practice (EBP)
  clinical situations, 54
  definition, 54, 58
  five-step process, 54
  research-practice gap, 54, 55
  systemic reviews, 54, 55

© Springer International Publishing Switzerland 2016
H. Mullins-Owens, *Integrative Health Services*,
SpringerBriefs in Public Health, DOI 10.1007/978-3-319-29857-3

Printed in the United States
By Bookmasters